Zero Point Energy
The Fuel of the Future

Thomas Valone, PhD, PE

Integrity Research Institute

Non-profit 501(c)3 organization

100% of the proceeds from this book go to benefit IRI

for Nikola Tesla, the true visionary who first defined useful ZPE

Acknowledgement is given to Hendrik B.G. Casimir, Hal Puthoff, Bernard Haisch, Umar Mohideen, Steven Lamoreaux, Federico Capasso, Eric Davis, Fabrizio Pinto, Robert Forward, Jordan Maclay, David Froning, Gene Bazan, Christian Beck, Miguel Alcubierre, R. S. Decca, David Iannuzzi, Aviation Week & Space Technology, New Scientist, Nature, and Scientific American for their pioneering efforts and contributions to this book.

Zero Point Energy: The Fuel of the Future

Thomas F. Valone, PhD, PE

First Edition, 2007
Second Edition, 2008
Third Edition, 2009

ISBN 978-0-9641070-2-1

Integrity Research Institute
5020 Sunnyside Avenue, Suite 209, Beltsville MD 20705
301-220-0440, 800-295-7674, 202-452-7674
www.IntegrityResearchInstitute.org
IRI@starpower.net

Email or write for an IRI publication catalog

Table of Contents

Special Note for the Third Edition

Since the publication of this book in 2007, McGraw-Hill publishers has picked up Chapter 1 for reprinting in a college textbook,

Taking Sides: Energy and Society
Editor(s): Thomas A. Easton
Publication Date: 12/5/2008

A lecture of mine on zero point energy has also ended up on Google video.

Furthermore, there have recently been reports of the independent verification of the predictions in Chapter 4. More details are contained in the author's 2009 paper, "Proposed Use of Zero Bias Diode Arrays as Thermal Electric Noise Rectifiers and Non-Thermal Energy Harvesters" published by the American Institute of Physics in the *Proceedings of Space, Propulsion and Energy Sciences Internationals Forum, Workshop on Future Energy Sources* (see Appendix).

Preface

Overview

When we look at a scientific revolution, there are always individuals of stellar ability and performance who lead the way intellectually, besides those who provide resistance.[1] Drs. H.B.G. Casimir, Timothy Boyer and Robert Forward were a few of the earliest pioneers in the zero-point energy arena. As can be seen from the graph below,[2] Casimir's original paper, about the zero-point energy force named after him, has had a worldwide, exponential effect on increasing the number of Casimir papers, showing the rapidly growing interest in the topic.

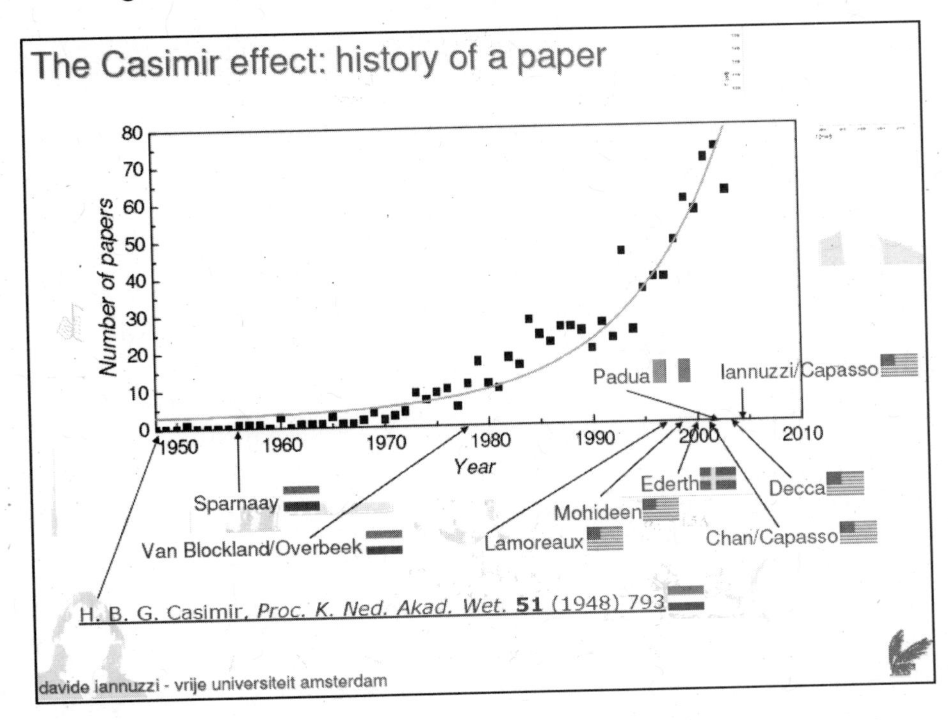

Dr. Boyer's articles in *Scientific American* and *Physical Review* spawned additional interest in fundamental processes and a new theory called "Stochastic Electrodynamics" (SED) which has provided answers to questions that quantum mechanics could not. Dr. Forward's articles started everyone thinking about possibly converting the Casimir force to usable electricity, which had not been considered before. Dr. Fabrizio Pinto, who has inspired me the most, is one of the more recent, amazing physicists to publish and patented his ZPE inventions. Dr. Peter Milonni's textbook, *The Quantum Vacuum*, provided the impetus for me and others to take the field of ZPE seriously. Even Tom Grace's novel, *Quantum*, a thriller centered on a free energy device, provided the intrigue and suspense for me to pursue this topic more fervently.[3] However, it was Nikola Tesla's prediction of the field of zero point energy that inspired me with his contagious prophetic conviction.

After completing my "Feasibility Study of Zero-Point Energy Extraction from the Quantum Vacuum for the Performance of Useful Work" which was also used in partial satisfaction for a PhD in General Engineering, it became apparent that the discoveries that I reported on were not accessible to the general public because of the technical nature of the report and the physics equations which the study included. (The study has been released as a book, with a new title, ***Practical Conversion of Zero-Point Energy***, available from all major book distributors, including Amazon.com.)

Easy Reading

Upon designing the difficulty level of this book, it has been of paramount importance to me to make the book easy to read. Granted the topic is not an easy one but as a former community college teacher (Erie Community College, Buffalo, NY), I became an expert at reducing physics and electrical engineering to a high school level for general consumption. Adding personal insights, humor and sensory data, I worked the entertainment value of the book so that it could be high enough to keep your attention. Even so, my hope is that each reader will grasp the significance of the true energy revolution that this book heralds.

Exciting Future

Ever since 1980, my avocation has been directed toward "future energy" with annual lectures and a paper on the topic. Our nonprofit organization, Integrity Research Institute, formed in 1990, offers a free, monthly email news service called "Future Energy eNews," and has my lectures, CDs, DVDs, books and papers on "Future Energy Technologies" available to the public. You may also request a free sample DVD of a 2006 lecture on this topic, which includes zero-point energy, given at the Conference on Future Energy (COFE) in Washington DC, just by visiting the order page at www.IntegrityResearchInstitute.org. All of this work has kept me abreast of the developments in future energy, which in my opinion, are headed toward a compact, portable electricity generator that consumes no fossil fuel but produces useful amounts of electrical power.

Our basic survival as a species depends on such an invention as we quickly exhaust all other energy sources. Surprisingly, many battery, ultracapacitor, and compact fuel cell developments fit the bill already. To gain a perspective on our dire situation as an evolving human race, just think of any major disaster where the concomitant deprivation always includes the loss of electricity. Looking at vast territories on earth that are uninhabited due to a lack of resources. Survey the underprivileged, third world countries that don't have even their basic needs supplied. Think of space travel for any appreciable distance, such as to Mars. All of these situations demand a denser, high quality electrical generator that is longer lasting, preferably on the order of years. Where there might not be sun, wind, or thermal energy, zero point energy fuel cells will be the only sustainable alternative in the future.

Zero Point Energy

The term, zero-point energy, has been written with and without a hyphen (e.g., zero point energy) in all of the major journals. Therefore, in this book, you will see it spelled both ways. It is also important for me to express to you my sincere desire for the truth in this field and every other one. I tend to follow the Miller and Morris, ***Fourth Generation R & D*** approach to new technology, which has specific concepts for recognizing a major trend of the future.[4] *Zero point energy development qualifies as a major trend.* All of the evidence in this book and the discoveries yet to come point to this truth. Esoteric religious mystics also find it to be the Holy Grail of energy, which indicates a fundamental drive in the collective unconscious for fulfillment of an inherent need. In fact, one "history of energy" text that I read to prepare for this project indicated that the development of every society depends solely upon developing higher quality energy sources. Furthermore, the collapse of civilizations, they claimed, was due to the stagnation of the energy source depended upon and a lack of a superior replacement in time.

Resistance to Looking Under the Rock

There is a problem. Many scientists in the conventional academic world represent the status quo, which they seek to maintain. Anything that remotely looks like perpetual motion or free energy is summarily rejected out of hand by this group, even if it can be demonstrated. For example, Professor John Barrow from Cambridge University insists that,

> In the last few years a public controversy has arisen as to whether it is possible to extract and utilise the zero-point vacuum energy as a source of energy. A small group of physicists, led by American physicist Harold Puthoff have claimed that we can tap into the infinite sea of zero-point fluctuations. They have so far failed to convince others that the zero-point energy is available to us in any sense. This is a modern version of the old quest for a perpetual motion

> machine: a source of potentially unlimited clean energy, at no cost....The consensus is that things are far less spectacular. It is hard to see how we could usefully extract zero-point energy. It defines the minimum energy that an atom could possess. If we were able to extract some of it the atom would need to end up in an even lower energy state, which is simply not available.[5]

This narrow-minded opinion proves that converting or extracting zero-point energy for useful work is still plagued by ignorance, prejudice and disbelief. The majority of physicists do not in general acknowledge the emerging opportunities from fundamental discoveries of zero-point energy. Instead, there are many expositions from prominent authorities explaining why the use of ZPE is forbidden and bordering on the irrational. A scientific editorial opinion states,

> Exactly how much 'zero-point energy' resides in the vacuum is unknown. Some cosmologists have speculated that at the beginning of the universe, when conditions everywhere were more like those inside a black hole, vacuum energy was high and may have even triggered the big bang. Today the energy level should be lower. But to a few optimists, a rich supply still awaits if only we knew how to tap into it. These maverick proponents have postulated that the zero-point energy could explain 'cold fusion,' inertia, and other phenomena and might someday serve as part of a 'negative mass' system for propelling spacecraft.[6]

With convincing skeptical arguments like these from the "experts," how can the extraction of ZPE for the performance of useful work ever be considered reasonable? What engineering protocol can be theoretically developed for the extraction of ZPE if it can be reasonably considered to be feasible? These are some issues that are points of discussion for raising ZPE consciousness.

Sustainable Living

As we move into the 21st century, it is an honor to help usher in a vital ingredient for the future trials and tribulations that will inevitably assault this world. When we consider the placid, past ten thousand years of inter-glacial period with relatively even temperature and constant sea level, it is important to realize that the earth is normally in an Ice Age for 90% of the time (lasts about ninety thousand years) and should be entering one right now. A recent *Scientific American* cover story indicated that it is only due to human activity that we haven't already entered the next Ice Age. However, humans have now overdone the compensating "thermal forcing" of the planet. Furthermore, all of the volcanoes, calderas, landslides, earthquakes, tsunamis, hurricanes, and most dangerous of all, meteorites, scheduled to become global catastrophes, at a time when the world population density is the highest it's ever been in urban areas, will necessarily cause a large-scale loss of life.

What a blessing it will be for everyone to have a small sealed box in his or her house and place of business that supplies a sustainable amount of electricity with a zero-point energy generator. Cooking, heating, air conditioning, water distillation, communication, computer and lighting needs will all be sustained during a major crisis, instead of being interrupted, as is the problem today. In his movie "Inconvenient Truth," Al Gore for example, cites drought brought on by climate change as a major cause of the bloodshed in Darfur. ZPE power sources and compatible electromagnetic medicine will help underdeveloped countries as well as the modern ones. With a source of electricity available everywhere, many consumable pharmaceutical drugs in Africa, which demand regular donations to sustain them, will slowly be replaced by a single, one-time investment of ZPE electrically-powered therapy devices (see the book, *Bioelectromagnetic Healing* by this author). This will be a lasting improvement to health and well-being worldwide.

I saved the technical, physics stuff for Chapters 9 and 10. Some readers may enjoy getting more of the scientific details that the last chapters provide. Hopefully, many of the graduates of this book will want to go onto the advanced text, *Practical Conversion of Zero-Point Energy*, that gives more of the real physics behind ZPE effects.

Private investment capital is needed for many of these projects to be completed. In the meantime, the rest of us can dream about the future that we all deserve to see materialized in our lifetime. Our mother earth needs the relief of clean zero

point energy so that she may heal the climate for our benefit. This is in keeping with the Gaia Principle. For almost three decades I have fought for the development of new sources of energy to avert the coming oil crisis, such as when I was on CNN on June 25, 2002. Hopefully, this book will help in that regard. To aid the understanding of the concepts and implications of this book much more easily, a one-hour illustrated lecture video is online which was very well received: http://video.google.com/videoplay?docid=-5738531568036565057

Thomas Valone
Washington DC

Figure 1.1 The first "Zero-Point Energy" toy on the market, a companion to the movie "The Incredibles," released in 2004

Chapter 1

Introduction to Zero Point Energy

Overview

Zero point energy is the sea of energy that pervades all of space, often called by scientists, "the physical vacuum." Perhaps a realization of the old ***ether theory*** or the Biblical *firmament*, it just happens to be the biggest sea of energy that is known to exist. Not only is it big but its energy is estimated to exceed nuclear energy densities. Even a small piece of it is "worth its weight in gold." What is it? It is "the kinetic energy retained by the molecules of a substance at a temperature of absolute zero."[8] Still, most people are not sure what this "zero point energy" (ZPE) is and whether it can be useful for human energy needs.

> **Ether (Aether) Theory.** Light may use something as a medium, similar to how air and water is used by sound. Michelson & Morley proved it didn't exist but their experiment has been countered by Silvertooth[7] and others with improved accuracy.

Does it offer a source of unlimited energy for homes, cars, and space travel? Depending on whom we talk to, ZPE can do everything and ZPE can do nothing useful. How can the energy be converted to produce electricity? It may be our primitive 20th century upbringing that stops us from putting a paddlewheel in this sea. What is the basic explanation of ZPE? *Space is quantized and virtual particles abound.* What are the new discoveries that have rocked the U.S. Patent Office, NASA, *Physical Review*, *Scientific American*, *Discover*, *New Scientist*, and the *New York Times*? What are some of the ZPE concepts that we should know about? These are the question that this book will answer in the following chapters. This is a very useful chapter and every new term is defined in a special definition box for your convenience.

What is Zero Point Energy?

> **Virtual Particle.** Also called a "virtual quantum," it is an intermediate state where energy is not conserved. The theory of exchange of coulomb energy between two electrons involves an emission of virtual quanta by one and the absorption by the other. Virtual particles abound in the vacuum.

Maybe ZPE can shoot from a gun, like in the movie "The Incredibles" (Fig. 1.1)?[9] Some scientists like to talk about the vast field of zero point energy pervading all space, as the *zero point field* (ZPF). We can envision the ZPF as a big sea in which we are all submerged. Contrasted with that is ZPE that locally involves energetic stuff on a microscale, which we can measure. Thus, ZPE is the energy that comprises the ZPF. Dr. Fred Wolf explains:

> No matter how cool we make the chamber as we compress the gas, we would find that we could no obtain total order. Greater confinement of each molecule would produce, according to the uncertainty principle, a greater uncertainty in its possible speed and therefore less certainty about its individual behavior. The gas would exhibit what is called zero-point energy. Even though its temperature was reduced to absolute zero, the molecules would still continue to move. Each molecule, however, would no longer be able to occupy a single position at a single time. Instead each would 'spread out' throughout the whole volume of the chamber.[10]

With the discovery of ZPE, scientists find that space is rich with activity from ***virtual particles*** and full of energy. Therefore, physicists like to call it the "physical vacuum" when they want to talk about ZPE. Furthermore, the vacuum also vibrates and "fluctuates." *In fact, that is the very essence of ZPE.*

Vacuum fluctuations are even predicted by a branch of physics, started by Albert Einstein, Neils Bohr, and Werner Heisenberg, called *quantum mechanics*. "Vacuum fluctuations" will be regarded as the same thing as ZPE, which are "a disturbance in the Force, Luke."

Another aspect of both ZPE and ZPF is that the "vacuum" was supposed to be empty. This is the only "leap of faith" that is required of the reader: to keep an open mind to the fact that theory and experiment agrees that the vacuum is not empty. Instead, it is full of activity and, most importantly, it can spill over into the real world. As explained in later chapters, classical physics predicts the presence of zero point energy. The way it was discovered involved emptying a container of everything including the matter, gas, and any heat energy. The only thing left in that container, as the heat energy approaches the absolute *zero point* (0°K) will be the vacuum itself, assuming electromagnetism and gravity play no part. This is why it is called *zero point* energy. (More historical ZPE information in Chapter 2.)

Scientists have cooled specimens to less than 1°K of the absolute zero point of temperature. A famous experiment proving the existence of ZPE involves cooling helium to within microdegrees of absolute zero temperature (between –272C and absolute the zero point of –273C). Amazingly, it will still remain a liquid! Only ZPE can account for the source of energy that is preventing helium from freezing.

Besides the classical explanation of zero-point energy referred to above, there are rigorous derivations from quantum physics that prove its existence. As one quantum mechanics text states, "It is possible to get a fair estimate of the zero point energy using the uncertainty

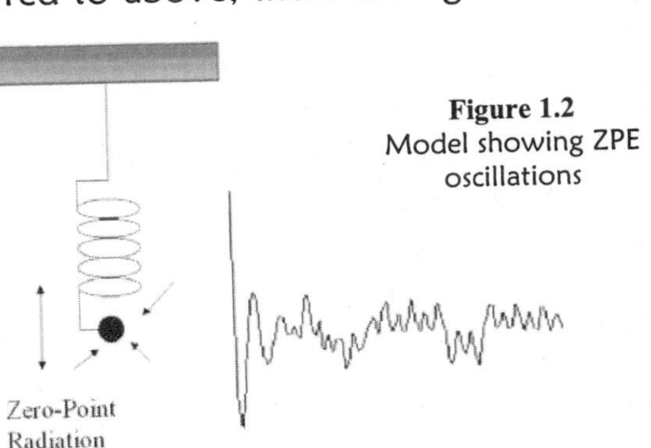

Figure 1.2
Model showing ZPE oscillations

principle alone" (see Appendix). Because Heisenberg's uncertainty principle is so simple and fundamental, this implies that ZPE is also the same: simple and fundamental.

Infinite Energy

Everything about zero point energy is amazing. This apparently keeps physicists in a state of incredulity, unable to grasp its significance. For example, there is the question of whether the ZPF is *conservative*. (Conservative fields in physics are those that *conserve* and don't create or destroy energy, thus obeying the First Law of Thermodynamics.) If ZPE is not conservative, then we can extract "an infinite amount of energy" from the vacuum, according to Dr. Robert Forward. However, if it is proven that the ZPF is a conservative field, we can still extract energy from it. It is just that we would have to put energy in and store it somehow to get it out again.

The evidence seems to favor a **nonconservative ZPF** so far, with an **open system** bath or plenum. In addition, the capability of ZPF storage and retrieval has convincingly been presented by Dr. Forward, as presented in the next chapters.[11] An article entitled, "Energy Unlimited" appeared a few years ago when Professor Jordan Maclay received a NASA grant to try to extract ZPE from elongated, oscillating, tiny metal boxes.[12] A physics journal article points out, "However, vacuum fluctuations remain a matter of debate, mainly because their energy is infinite. More strikingly, **their energy per unit volume is infinite.**"[13] Interestingly, this seems to create intellectual difficulties which can only be artificially eliminated. "Problems with the infinite energy of vacuum fluctuations has led to the view that vacuum energy may be forced to vanish by definition...[from] the need to regularize the infinite energy-momentum tensor." However, this causes even more complications: "This procedure gives rise to ambiguities and anomalies, that lead to a breakdown at the quantum level of usual symmetry properties of the energy-momentum tensor."[14]

Some physicists defend the infinite energy theory because presently there is no known limit to how small an electromagnetic vibration can be.[15] Therefore, they argue that there has to be infinite possible electromagnetic vibrations in the ZPF. This logically leads to the conclusion that infinite vibrations yield infinite energy content. However, this argument seems to apply more to energy density rather than total energy content. As Dr. Milonni explains, **"The zero-point energy of the vacuum is infinite in any finite volume."**[16] "A charged particle in the vacuum will therefore always see a zero-point field of infinite energy density"[17] which supports the unlimited view of the quantum vacuum plenum.

For example, imagine what is the smallest vibration that could exist. That tiny wavelength has to resonate with a correspondingly high frequency.[18] This calculation still leads to very high energy density and a really big number for the total ZPE available in the universe. We will call this the "limited" ZPF as opposed to the "unlimited" ZPF that yields infinite energy density.[19] (More details are in ref. 18 and in later chapters.)

> In an interview taped for television on PBS's *Scientific American Frontiers*, which aired in November (1997), Harold E. Puthoff, the director of the Institute for Advanced Studies, observed: **'For the chauvinists in the field like ourselves, we think the 21st century could be the zero-point-energy age.'** That conceit is not shared by the majority of physicists; some even regard such optimism as pseudoscience that could leech funds from legitimate research. The conventional view is that the energy in the vacuum is miniscule.[20]

Ten years later, this skeptical viewpoint is unfortunately still held by the physicists/scientists who want to keep their funding.

Contrary to this pessimistic, irrational belief, the actual scientific estimate of energy density of even the *limited ZPF* (bounded by a maximum frequency) is astounding. It is much more than we humans normally can comprehend. For example, if we presume that the minimum possible wavelength is limited to the size of the proton, the famous Nobel Prize winning physicist, Richard Feynman, calculated that the energy density of the ZPF would be ten raised to the 108th power ***joules*** per cubic centimeter (10^{108}J/cc). Today, physicists want to look at even smaller vibration units like subatomic particles, etc. This makes the ZPF energy density escalate even more. Just as a comparison, if we converted energy to mass using $E=mc^2$, we find that the equivalent "mass density" of the ZPF is ten to the 94th power grams per cubic centimeter (10^{94} g/cc). Compare that with typical nuclear densities of ten to the 14th power grams per cubic centimeter (10^{14} g/cc).[21] Therefore, gram for gram, *ZPE offers almost ten to the eightieth times more energy for the same amount of space than nuclear reactors.* Therefore, if we presume similar energy conversion efficiency, then 1 ZPE Engine = 10^{80} Nukes. This leads to the surprising ZPE conclusion: ***Space itself contains more energy than matter does for any given volume.***

Free Energy

Fig. 1.3 This ten-year old videov (DVD) is a classic

This raises the exciting and controversial issue of "free energy." This is "**one of the world's twenty greatest unsolved problems**," according to a new book published by Prentice Hall that devotes an entire chapter to free energy.[22] Historically, when natural gas (1950s) and again when nuclear power (1960s) was introduced to society, "cheap energy" was their advertising slogan. Back in 1903, Nikola Tesla completed the Wardenclyffe Tower on Long Island, in order to broadcast cheap electrical energy to Europe. Tesla was stopped by J.P. Morgan who wanted to know how he could possibly put a meter on it. This caused Tesla's free energy dream to be suppressed. Today this movement is still a suppressed, popular conviction held by the majority opposed to the high cost of electrical grid power and ridiculed by mainstream physics. In 1995, I was the technical consultant to a bold, pioneering video "**Free Energy: Race to Zero Point**" which introduced and discussed ZPE with state of the art graphics and professional narrator. It also contains examples of promising inventions that showed characteristics of self-powered operation, though no endorsement was made of their outcome.[23]

Now with the advent of ZPE, localized free energy looks much more promising than ever before. In 2001, *Popular Mechanics* featured an article talking about putting "free energy to work" moving a nanoscale seesaw (see Fig. 9.1).[24] In 1998, a physicist with the Jet Propulsion Laboratory invented a ***nanoscale*** ZPE engine that pumps electrons, with the help of a tiny laser, just like an electrical generator. Dr. Pinto states,

> In the event of no other alternative explanations, one should conclude that major technological advances in the area of endless, by-product free-energy production could be achieved.[25]

Pinto's accomplishment has brought much needed legitimacy to the ZPE conversion arena. Another exciting endorsement has come from the prophetic Arthur C. Clarke who was recently overheard talking to Astronaut Buzz Aldrin in Sri Lanka, broaching the issue of zero point energy,

> I'm now convinced that there are new forms of energy, which we are tapping, and they make even nuclear energy look trivial in comparison. And when we control those energy sources, the universe will open up.[26]

Nanotechnology. Atomic size engineering, on the scale of a billionth of a meter (10^{-9} m), which is a nanometer. Such technology often uses scanning tunneling microscopes and other means to position even single atoms. Nanoscale motors, nanotubes, and nanobots are a few of the creations of this exotic world.

The Need for ZPEED?

In the movie "Top Gun" with Tom Cruise, there is a famous line: "I feel the need for speed." Today the need for a nonpolluting, abundant energy source like ZPE is greater than ever. Maybe it will be called "zpeed" in the future for that reason, with milder addictive side-effects than we have now. Our world is battling another energy crisis of unprecedented proportions. However, the present levels of greenhouse gases that are by-products of fossil fuel energy are already impacting the climate and weather. The environmentalists' demand to reduce the carbon emissions by 60% to 80%, in order to stabilize the earth's atmosphere, clashes sharply with the public's increasing demand for energy. While the world relies heavily on an oil-based energy that experts say is at peak production, new futuristic energy sources are still out of reach.

Fortunately, the ultimate goal of ZPE vacuum engineers is none other than free energy (with a single capital investment) and unlimited amounts of electrical power. Zero point energy is the much-anticipated promise of the future. It is the omnipresent bulwark of nature's machinery, and the most abundant energy source in the universe. It already powers a surprising number of processes from quasars to atoms, while also linked by theoretical physicists to inertia and gravity. The applications of ZPE are limitless. We just need to design effective transducers to put the energy to use.

Dr. Marc Millis, at NASA's Breakthrough Propulsion Physics Program, calls zero point energy the "leading candidate" for interstellar travel. It also is much more vital for interplanetary travel than NASA's overgrown firecrackers, invented by the Chinese over one thousand years ago.

Facing the future with knowledge and forethought will empower us to find success in the midst of our dilemma. The questions are simply:

(1) How can energy be used anywhere without burning something?

(2) How can we travel quickly in space without exploding something?

Zero point energy is the only long-term solution to these questions.

The Vacuum Is Not Empty

Let's look more closely at the so-called vacuum of empty space. Surprisingly, physics tells us that it's pretty much the same thing as what is inside your body, this book, and everything around you. All ordinary terrestrial matter is really 99.99% "empty" space. As proof, let the nucleus of any atom expand to the size of a football placed on the 50-yard line of an American football field (during half-time of course). The first electron shell would be somewhere in the upper balcony with just a speck of dust flying around the field. When I first heard this story from Dr. Fritjof Capra, the author of *The Tao of Physics*, I was instantly overwhelmed by a feeling of emptiness. *Somehow, physical reality supports the spider web of matter only with fields and forces.*

The other guided imagery that Dr. Capra liked to use was one designed to give everyone a relative appreciation for the size of atoms. He said to take an orange and blow it up to the size of the earth in your mind. Once you do that, each atom in the orange is about the size of a cherry. He said the proportions are fairly accurate. So the image we are left with is an orange the size of the earth filled with closely packed cherries.

While presenting his keynote address at our institute's Conference on Future Energy in 2006 (available on DVD), Dr. Fabrizio Pinto, formerly with Jet Propulsion Lab, indicated that **the vacuum has "pressure, density, and substance."** He also emphasized the fact that it can move physical objects under certain conditions. Pinto provoked the issue by asking, "Is empty space really worth nothing?"[27] This became a rhetorical question in the face of all of the inventions that Pinto subsequently presented to tap the ZPE for energy and propulsion applications. His best known is the Casimir Engine that is reviewed in detail elsewhere in this book. It has an incontrovertible thermodynamic engine cycle described in his *Physical Review* journal article as well as patent protection.[28]

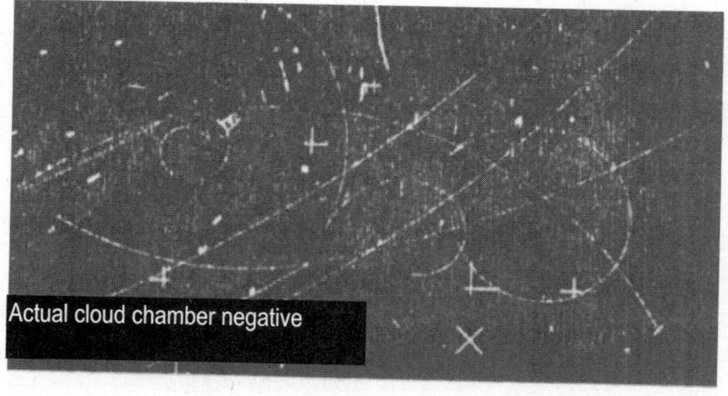

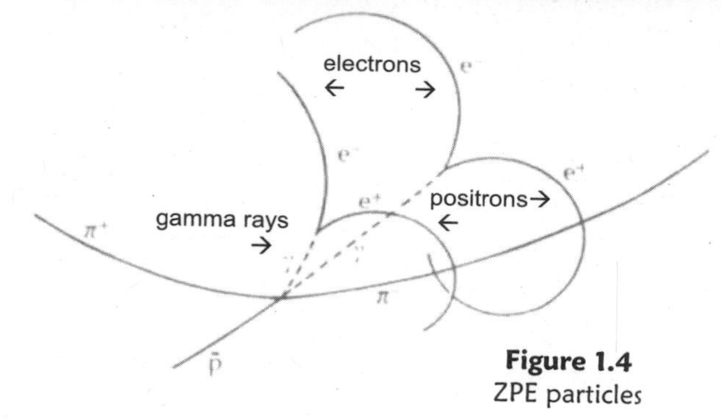

Figure 1.4
ZPE particles

Quantum electrodynamics (QED) predicts that the vacuum spawns particles that spontaneously pop in and out of existence. Their time of existence is strictly limited by the ***uncertainty principle*** but they create some havoc while they bounce around during their brief lifespan, from virtual existence to real existence and back. The churning "quantum foam," as it is popularly called, is believed to extend throughout the universe even filling the empty space within the atoms in human bodies. Furthermore, theory predicts that the electron levels in each atom are sustained and supported by quantum fluctuations, which tend to *polarize* (become more manifested in a certain locality) near the boundaries of charged particles, like a proton or electron. This is explained in the later section in this chapter entitled, "Why Atoms Don't Collapse." Dr. Pinto explained during his speech that without ZPE, all of our atoms in our bodies would collapse into a small pile on the floor.[29] With that provocative image, it is no wonder that many physicists are eagerly learning more about such a fundamental field of energy.

When I first learned about Dr. Paul Dirac's theory of the *positron* (the antimatter version of the electron that has a positive charge). Dirac, a Nobel Prize winner, theorized in 1928 that virtual, negative energy states of electrons are not normally observable.

Ucertainty Principle. Postulated by Werner Heisenberg, it disturbed a lot of people especially Einstein who countered with "God does not play dice with the universe." Specifically, it limits the precise, simultaneous measurement of distance and momentum ($\Delta x \bullet \Delta p \geq h/4\pi$) or energy and time ($\Delta E \bullet \Delta t \geq h/4\pi$). ($\Delta$ is the uncertainty) so that their product has to be bigger than Planck's constant divided by 4π.

However, given sufficient energy, about 1.2 million electron volts (1.2 MeV), an electron can be knocked out of its negative energy vacuum state, leaving a "hole." The strangest thought in physics probably occurred when Dirac, in a perfectly sober state of mind, wrote in 1928 about how this "hole" could behave like a real particle, only with opposite charge and mass. This hole was subsequently found to actually be a particle and was named a "positron." In 1977, Dirac reminisced about those days, giving us a unique view of the ZPF, saying,

> Now, with quantum mechanics, we cannot exclude transitions from positive energy states to negative states, and that means that we cannot exclude negative energy states from our theory. If we cannot exclude them, we must find a method of physical interpretation by adopting a new picture of the vacuum. Previously, people have thought of the vacuum as a region of space that is completely empty, a region of space that does not contain anything at all. Now we must adopt a new picture. We may say that the vacuum is a region of space where we have the lowest possible energy. Now, to get the lowest energy we must fill up all the states of negative energy. The more electrons we can put into states of negative energy, the lower the total energy becomes, because each electron in a state of negative energy means a reduction in the total energy. Thus we must set up a new picture of the vacuum in which all the negative

energy states are occupied and all the positive energy states are unoccupied.[30] (see Fig. 1.5)

When a gamma ray decays to a positron and an electron in Wilson cloud chamber experiments (Figure 1.4), the tracks left by the particle and antiparticle are clearly seen as they move away from each other.[31] (The spirals are caused by the interaction with an external magnetic field.) Yet, Dirac explains that in those days, people were very reluctant to postulate the existence of a new particle, even though they had seen it in the lab.

When asked, "Why had experimenters never observed these antiparticles in cloud chambers?" Dirac says, "I think the only answer to that question is that they were prejudiced against new particles...*They had never observed positrons, because they really turned a blind eye to them when they had evidence for them.*"[32]

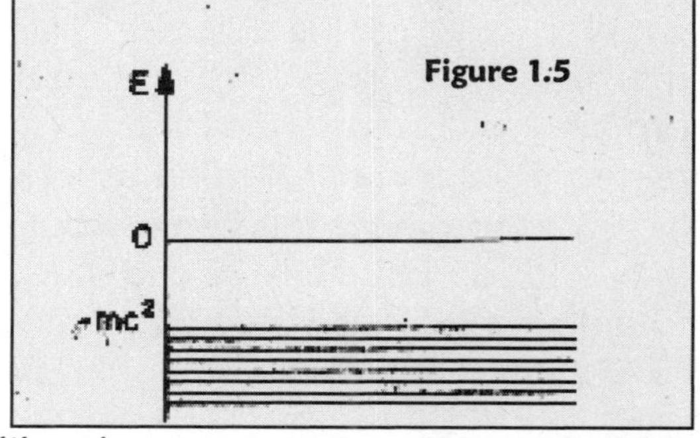

Figure 1.5

This is very much like the present situation with ZPE. Physicists acknowledge the theory, publish papers on the subject, and still do nothing practical to solve the energy crisis. Even today, with numerous experiments verifying the theory of ZPE and many proposed ZPE engine cycles, only a few scientists are actively researching ZPE converters. Many still believe that energy cannot be extracted from virtual processes, though the energy density is the highest we can imagine. Therefore, I predict that future job openings for a "quantum mechanic" or a "vacuum engineer" are sure to be lucrative though and very rewarding in my opinion.

Ironically, history is filled with examples of reluctance to use new discoveries. Take the random heat energy that only applies pressure. That is thermal energy, which may never have been harnessed into a steam engine, a heat pump, or an automobile, if the skeptics had won out. Even more exasperating is the story about the Wright brothers' attempt to convince anyone, including the Patent Office, that their airplane actually flew. How about Edison's battle against Nikola Tesla, proving that AC electricity is far too dangerous, by electrocuting dogs in public at state fairs? Will rampant skepticism and prejudice also delay the research, development and mass marketing of ZPE converters?

The Shape of Nothing

Now let's stretch our imagination to the smallest level possible and see what is going on. The diagram representing "The Shape of Nothing" (Figure 1.6) is an artist's rendition of quantum foam. It is subatomic, showing virtual elementary particles popping in and out of existence.[33] So we have to think really small when we're thinking about zero point energy. Physicists theorize that on an infinitesimally small scale, far, far smaller than the diameter of atomic nucleus, quantum fluctuations produce a foam of erupting and collapsing, virtual

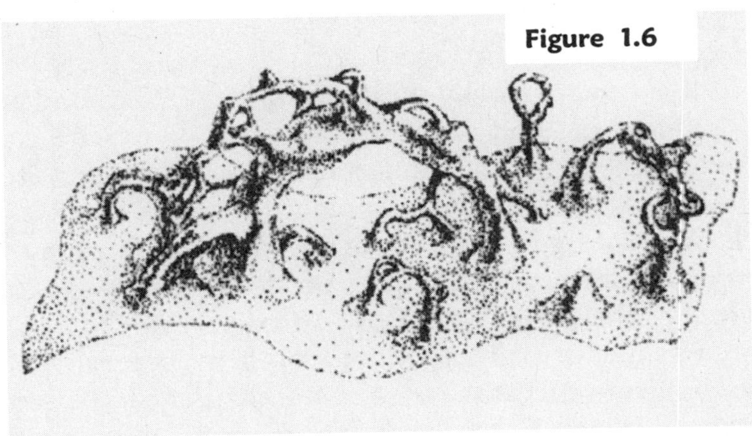
Figure 1.6

particles, visualized as a topographic distortion of the fabric of spacetime. Today, physicists propose that even on a level that is many times smaller than a proton, quantum foam is churning out particles, mostly electrons and positrons, as we will find in later chapters.

On the cover of *Science* magazine, "The Breakthrough of the Year" article (see Fig. 10.1) blamed zero point energy for our "antigravity" universe whose expansion is now found to be accelerating.[34] In a Los Alamos Laboratory, the Casimir force from ZPE was recently measured (1997) for the first time with conductive plates and found to be within 5% agreement with theory.[35] At AT&T Bell Labs, a specially designed semiconductor device will only resonantly tunnel because of the presence of zero point energy.[36] In the *Washington Times*, zero point energy was used to explain the advanced antigravity propulsion systems the government supposedly has in classified programs, to "produce vast amounts of energy without any pollution."[37] At Edwards Air Force Base, scientists received a patent for a ZPE converter that can optimally use subatomic particles as "ZPE electromagnetic receivers" to produce useful electricity (see later chapters). At NASA's Breakthrough Propulsion Physics project, a three-year grant was awarded for the first time, to extract ZPE for useful purposes.[38]

Why Atoms Don't Collapse

It is now generally understood that ZPE is fundamentally responsible for atomic stability. When I spoke with Dr. Harold Puthoff, from the Institute for Advanced Studies in Austin, Texas, he described his plans for measuring the electron levels of a hydrogen atom while it is in a Casimir cavity (we will return to this topic later). He proposed years ago that ZPE determines the ground state of hydrogen (see Ch. 2). Now he finally has access to the facilities to test his theory in a Casimir cavity which restricts the range of frequencies of ZPE around the hydrogen atoms.[39] According to Puthoff, this should show depressed or lowered electron level energy as a result of less vacuum fluctuation activity that normally supports each electron level. Perhaps Dr. Randell Mills' "hydrinos" still have a second lease on life.[40]

The *New Scientist* journal, with the same article title as this section, talks about ZPE and describes it as the vacuum spontaneously, in not violating the uncertainty principle, giving rise to particles: The quantum vacuum does this continuously thus creating the Casimir force.

> Thus there is a dynamic equilibrium in which the zero-point energy stabilizes the electron in a set ground-state orbit. It seems that the very stability of matter itself appears to depend on an underlying sea of electromagnetic zero-point energy.[41]

Virtual Particle Atmosphere

This constant virtual particle flux of the ZPE is especially noticeable near the boundaries of bigger particles, because the intense electric field gradient causes a more prodigious decay of the vacuum. This particular condition is aggravated by the size of the particle. If it is a truly massive, hypothetical atom with 173 protons in the nucleus were possible, the **decay of the vacuum will occur spontaneously**, giving birth to electrons and positrons.

> Under these conditions the electron and positron are not vacuum fluctuations but are real particles, which exist indefinitely and can be detected...The essential condition for the decay of the vacuum is the presence of an intense electric field.[42]

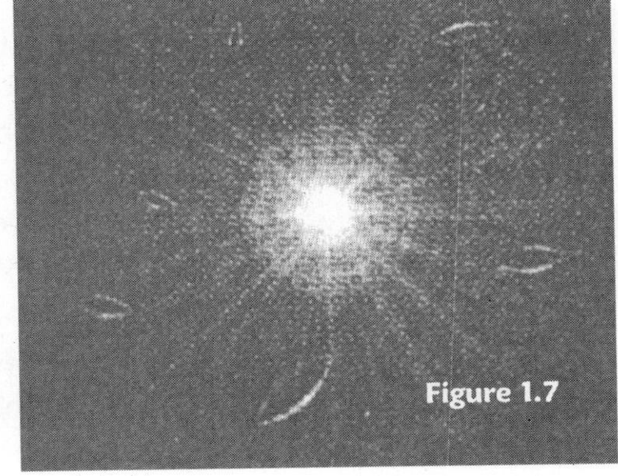
Figure 1.7

Since an electric field is measured in volts per meter (V/m), it will tend to be exponentially huge as we get extremely close to a charged particle. Therefore, the decay of the vacuum (called '**polarization**' of the vacuum) is potentially occurring all of the time near the boundary of charged particles, such as the nucleus and the electron. With this simple understanding, we can guess that the ZPE cloud surrounding the electron would tend to affect the bigger ZPE cloud surrounding the typical proton in some way.

> Since the 1930's, theorists have proposed that...virtual particles cloak the electron, in effect reducing the charge and electromagnetic force observed at a distance.[43]

Virtual particles are more likely to become real, or appear out of the vacuum, near the boundary of charged particles and even near the nuclei of atoms. These tend to "polarize" the vacuum. From the previous discussion, we know that is where the electric field is the most intense, approaching infinity. It is analogous to a tiny planet having an atmospheric shield that protects by absorbing energy from intruders. Likewise, every charged particle necessarily drags around with it a nasty vermin cloud of virtual particles, like buzzing flies, which cloak it from direct, outside perception and invasion.

In 1997, following Dirac's suggestion, Dr. Koltick performed an experiment using positrons to penetrate the virtual particle cloud surrounding the electron (Figure 1.6).[44] A particle accelerator at 58 billion electron volts (58×10^9 eV) enabled him to isolate the electron without creating other particles. He discovered a new value for the ***fine structure constant***. In reality, it was proven to be 1/128.5 instead of the smaller 1/137 that is traditionally observed for a fully screened electron. Therefore, the charge of the electron (e) must be bigger than physicists previously thought. Dr. Koltick literally has rewritten the physics textbooks with that one experiment and also proved to the skeptics how ZPE makes up a particle's "atmosphere." Such a polarized or non-polarized vacuum of virtual particles surrounds protons, neutrons and even micron-sized bacteria. In all cases, whether a

> **Fine Structure Constant.** Useful with the fine structure (splitting) of atomic energy levels seen in optical spectrum lines. It equals two times pi and the electron charge squared divided by Planck's constant and the speed of light ($2\pi e^2/hc$).

charged boundary or not, the particles apply pressure, called the Casimir force, to push the boundary inwards.

Summary

Talking about ZPE may seem to be bordering on fantasy for most people, irrespective of whether you are a physicist or not. All of us want to be assured that any new source of energy is real and will help to cook our eggs or run our cars. Otherwise, scientists tend to treat new concepts as a black hole where lots of theory goes in and nothing comes out.

In the following chapters, a whole universe of zero point energy facts and ideas will open up for the reader. So compelling is the information that I believe you will be challenged to take it seriously and plan your energy-abundant future accordingly. I predict that third world countries will finally have their basic needs met with small, portable energy units distributed everywhere. With that capability comes a necessary burden to use this new emerging energy source wisely and constructively for everyone's benefit.

My suggestion is for the civilian scientists and entrepreneurs to grab hold of the opportunity, share it with overseas colleagues, and get it to market, before the military declares it "classified." You may laugh but this is the best strategy, tested and proven by inventor Ken Shoulders. For a few years, his discovery of "**electron charge clusters**," with estimates of several times overunity, was ranked #2 on the National Critical Issues List, just under stealth bombers.[45] Ken told me that when the officer arrived at his door with the classification papers, he told him it was too late because he had already sent it overseas in the form of a few hardcover books. It worked so well that the officer was extremely angry as he had to retreat in defeat. To this day, electron charge cluster technology remains available for civilian applications due to Ken's heroism and concern for the common man.

Another example also demonstrates the keen interest the military has in new energy. As many remember, the original cold fusion experimental results were announced at a press conference at the University of Utah in 1989 by Pons and Fleischmann. Years later, Dr. Fleischmann told a small group of us at dinner, after a meeting in DC that on his way back to England he missed a connecting flight in San Francisco and unexpectedly had to check into a hotel room. As he opened the door to his room, the phone rang. The caller said, "This is Edward Teller. Don't hang up. I have a few questions for you." The first question he was asked was, "Can you make a bomb from cold fusion?" A strict pacifist, Fleischmann denied any possibility of a runaway, positive feedback event that could result in an explosion. However, he also told me that he wished the whole project had been classified so that he and Dr. Pons could have worked without press and public interference. The question of explosive potential has also been asked of ZPE, as suggested by Fig. 1.1, but to benefit the human race, for its ultimate survival and for the sake of the planet, this book emphasizes basic ZPE electricity and propulsion generators for public use rather than making more bombs for military use only.

For Further Information

There a number of popular books that introduce zero point energy with very readable details. For example, the recent book, *The Fabric of the Cosmos* by Brian Greene has a short section called "Quantum Jitters and Empty Space" with good graphics and discussion of the Casimir effect.[46] Also, *Einstein for Dummies* by Carolos Calle has a short section on zero point energy and the Casimir force.[47] The two-volumes on ZPE by Moray King, *Tapping the Zero Point Energy* and *Quest for Zero Point Energy*, are a fanciful survey of the literature, with lots of references and amazing ideas for conversion.[48] The book called *The Field* by Lynne McTaggart has a very interesting ZPE review in Chapter 2 called "The Sea of Light" that discusses the work of Puthoff, Shoulders, Rueda and Haisch in developing ZPE theories and conversion devices, much of which is based on actual conversations with them.[49] More details from *The Field* are reviewed later in this book (Ch. 8). The classic book, *The Secret of the Creative Vacuum* by John Davidson is a wonderful survey of free energy and ZPE with a chapter on "The Nature of Nothing."[50]

Science Times

The New York Times

TUESDAY, JANUARY 21, 1997 ← 1997

Physicists Confirm Power of Nothing, Measuring Force of Quantum 'Foam'

Fluctuations in the vacuum are the universal pulse of existence.

By MALCOLM W. BROWNE

FOR a half century, physicists have known that there is no such thing as absolute nothingness, and that the vacuum of empty space, devoid of even a single atom of matter, seethes with subtle activity. Now, with the help of a pair of metal plates and a fine wire, a scientist has directly measured the force exerted by fleeting fluctuations in the vacuum that pace the universal pulse of existence.

The sensitive experiment performed at the University of Washington in Seattle by Dr. Steve K. Lamoreaux, an atomic physicist who is now at Los Alamos National Laboratory, was described in a recent issue of the journal Physical Review Letters. Dr. Lamoreaux's results almost perfectly matched theoretical predictions based on quantum electrodynamics, a theory that touches on many of the riddles of existence and on the origin and fate of the universe.

The theory has been wonderfully accurate in predicting the results of subatomic particle experiments, and it has also been the basis of speculations verging on science fiction. One of the wilder ones is the possibility that the universal vacuum — the ubiquitous empty space of the universe — might actually be a false vacuum.

If that were so, something might cause the present-day universal vacuum to collapse to a different vacuum of a lower energy. The effect, propagating at the speed of light, would be the annihilation of all matter in the universe. There would be no warning for humankind; the earth and its inhabitants would simply cease to exist at

Continued on Page C6

Figure 2.1
The first confirmation of the Casimir force with conductive plates, performed at Los Alamos National Laboratories by Dr. Steven Lamoreaux in 1997

Chapter 2

History of Zero Point Energy

Overview

Zero-point energy (ZPE) is a universal natural phenomenon of great significance which has evolved from the historical development of ideas about the vacuum. Dr. Gary Johnson argues that it is also equivalent to the Biblical firmament.[51] The vacuum hypothesis was originally developed by the Greek philosophers Leucippus and Democritus.[52] In the 17th century, it was thought that a totally empty volume of space could be created by simply removing all gases. This was the first generally accepted concept of the vacuum. Late in the 19th century, however, it became apparent that the evacuated region still contained thermal radiation. To the natural philosophers of the day, it seemed that all of the remaining radiation might be eliminated by cooling.

Thus evolved the second concept of trying to achieve a really empty vacuum: cool it down to the zero point (degrees of temperature) after evacuation. The absolute zero temperature (-273C) was far removed from the technical possibilities of that century, so it seemed as if the problem was solved. In the 20th century, however, both theory and experiment have shown that there is a **non-thermal radiation in the vacuum** that persists even if the temperature could be lowered to absolute zero. (For example, helium has been cooled to within one degree of absolute zero but still remains a liquid.) This line of thinking explains the provocative name of "zero-point energy."

1891 Tesla was the first to describe essence of ZPE

World's Greatest Vacuum Engineers

In 1891, the visionary inventor and electrical futurist, Nikola Tesla, was the first to recognize the existence and properties of zero-point energy by stating,

> Throughout space there is energy. Is this energy static or kinetic? If static our hopes are in vain; if kinetic – and we know it is, for certain – then it is a mere question of time when men will succeed in attaching their machinery to the very wheelwork of Nature. Many generations may pass, but in time our machinery will be driven by a power obtainable at any point in the Universe.[53]

1912 Planck discovers ½hf

We need to credit Einstein's 1905 paper that introduced the "light-quanta" which he described as independent packs of energy.[54] Then in 1912, **Max Planck** published the first journal article to describe the discontinuous emission of radiation, based on the discrete quanta of energy.[55] In this paper, Planck's now-famous "blackbody" radiation equation contains the residual energy factor, *one half of a quantum*, as an additional term of ½hf, dependent on the frequency f, which is always greater than zero (where h = Planck's constant). It is therefore widely agreed that "**Planck's equation marked the birth of the concept of zero-point energy.**"[56] This mysterious factor was understood to signify the *average oscillator energy* still available to each field mode even when the temperature reaches absolute zero. In the meantime, Einstein published his "fluctuation formula" in 1909 which described the energy fluctuations of thermal radiation.[57] Today, "the particle term in the Einstein fluctuation formula is

1913 Einstein uses ZPE term

1928 Dirac posits positron

regarded as a consequence of zero-point field energy" but was not recognized as such.[58]

Within one year of its discovery, **Einstein**[59,60] and **Dirac**[61,62] saw the value of zero-point energy and promoted its fundamental importance. The 1913 paper by Einstein computed the specific heat of molecular hydrogen, *including zero-point energy*, which agreed very well with experiment. **Debye** also made calculations including zero-point energy (ZPE) and showed its effect on Roentgen ray (X-ray) diffraction.[63]

1914 Debye looks at X-rays

Throughout the next few decades, zero-point energy became intrinsically important to quantum mechanics with the birth of the uncertainty principle.

> "In 1927, **Heisenberg**, on the basis of the Einstein-de Broglie relations, showed that it is impossible to have a simultaneous knowledge of the [position] coordinate x and its conjugate momentum p to an arbitrary degree of accuracy, their uncertainties being given by the relation $\Delta x \, \Delta p \geq h / 4\pi$."[64]

1927 Heisenberg is very uncertain

By 1935, the application of harmonic oscillator models with various boundary conditions became a primary approach to quantum particle physics and atomic physics.[65] Quantum mechanics also evolved into "wave mechanics" and "matrix mechanics" which show two distinct methods for treating wave packets (particles). However, with the evolution of matrix mechanics came an intriguing application of matrix "operators" and "commutation relations" of x and p that today are well known in quantum mechanics. With these new tools, the "quantization of the harmonic oscillator" is all that is required to reveal the existence of the zero-point ground state energy.[66] To me this was a fascinating discovery. So I went back to my senior-year university quantum physics book to discover the derivation of the lowest energy state for ZPE. For those readers who are interested in how simple it is, I have added it to the Appendix.

> This residual energy is known as the zero-point energy, and is a direct consequence of the uncertainty principle. Basically, it is impossible to completely stop the motion of the oscillator, since if the motion were zero, the uncertainty in position Δx would be zero, resulting in an infinitely large uncertainty in momentum (since $\Delta p = h / 4\pi\Delta x$). The zero-point energy represents a sharing of the uncertainty in position and the uncertainty in momentum. The energy associated with the uncertainty in momentum gives the zero-point energy.[67]

Another important ingredient in the development of the understanding of zero-point energy came from the "Compton effect."

> Compelling confirmation of the concept of the photon as a concentrated bundle of energy was provided in 1923 by **A. H. Compton** who earned a Nobel prize for this work in 1927.[68]

1923 Compton scatters photons

Named after its discoverer, Compton scattering, as it is now known, can only be understood using the energy-frequency relation **E = hf** that was proposed previously by **Einstein** to explain the photoelectric effect in terms of Planck's constant, h.[69]

Ruminations about the zero-point vacuum field (ZPF), in conjunction with Einstein's famous equation **$E = mc^2$** and the limitations of the uncertainty principle, suggested that photons may also be created and destroyed "out of nothing." Such photons have been called "virtual" and are prohibited by classical laws of physics.

> But in quantum mechanics **the uncertainty principle allows energy conservation to be violated for a short time interval** $\Delta t = h / 4\pi\Delta E$. As long as the energy is conserved after this time, we can regard the virtual particle exchange as a small fluctuation of energy that is entirely consistent with quantum mechanics.[70]

1963 Feynmann diagrams ZPE

Such virtual particle exchanges later became an integral part of an advanced theory called quantum electrodynamics (QED) where "Feynmann diagrams," developed by Nobel-prize winner **Richard Feynmann** to describe particle collisions, often show the virtual photon exchange between the paths of two nearby particles.[71] Figure 2.2 shows a sample of the Compton scattering of a virtual photon as it contributes to the radiated energy effect of "bremsstrahlung" (see Glossary).[72]

Casimir Predicts a Measurable ZPE Effect

In 1948, it was predicted that virtual particle appearances should exert a force that is measurable.[73] Casimir not only predicted the presence of such a force but also explained why van der Waals forces dropped off unexpectedly at long range separation between atoms. The Casimir effect was first verified experimentally using a variety of conductive plates by Sparnaay.[74]

1948 Casimir predicts forces from ZPE

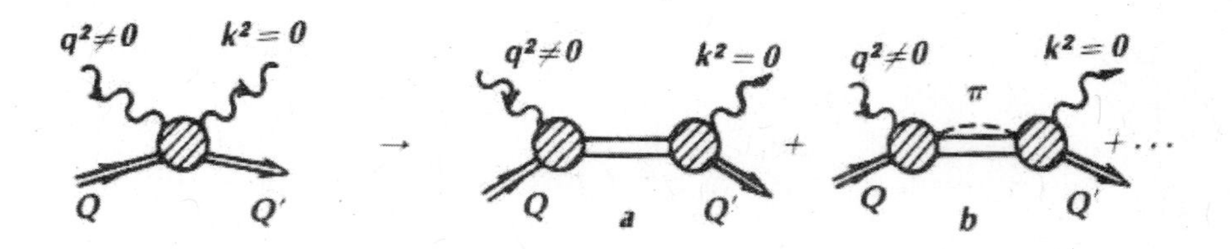

Figure 2.2 – A Feynmann diagram of a virtual photon hit a particle causing deflection (**Compton scattering**) seen as two separate processes (a) and (b)

There was still an interest for an improved test of the Casimir force using conductive plates as modeled in Casimir's paper to better accuracy than Sparnaay. In 1997, Dr. Lamoreaux, from Los Alamos Labs, performed the experiment with less than one micrometer (micron) spacing between gold-plated parallel plates attached to a torsion pendulum.[75] In retrospect, he found it to one of the most intellectually satisfying experiments that he ever performed since the results matched the theory so closely (within 5% accuracy). This event also elevated zero-point energy vacuum fluctuations to a higher level of public interest. **Even the *New York Times* covered the event** and published an easy-to-understand review of the historical breakthrough.[76]

1958 Sparnaay verifies Casimir

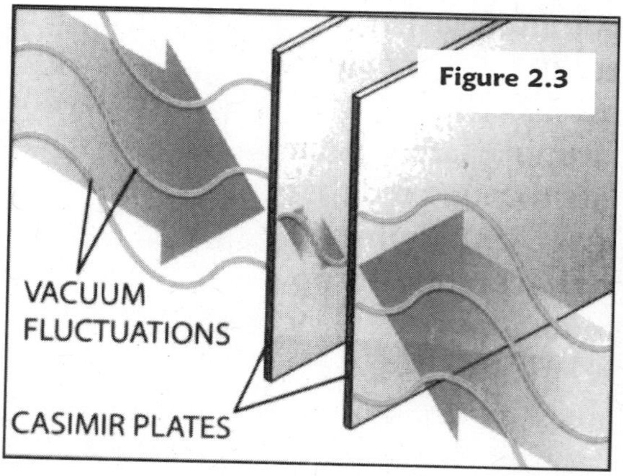

Figure 2.3

VACUUM FLUCTUATIONS

CASIMIR PLATES

The Casimir Effect has been posited as a force produced solely by activity in the empty vacuum (Fig. 2.3). The Casimir force (in pounds per square inch or newtons/m^2) is also <u>very powerful</u> at small distances (less than 1 micron or millionth of a meter). In nanotechnology, it causes "stiction" which makes tiny surfaces impossible to separate non-destructively. Besides being independent of temperature, it is inversely proportional to the fourth power (quartic) of the distance between the plates at larger distances and inversely proportional to the third power (cubic) of the distance between the plates at short distances.[77] Its frequency dependence is also a third power, meaning smaller is better!

Lamoreaux's results come as no surprise to anyone familiar with quantum electrodynamics, but they serve as a material confirmation of a bazaar theoretical prediction: that QED predicts the all-pervading vacuum continuously spawns particles and waves that spontaneously pop in and out of existence. Their time of existence is strictly limited by the uncertainty principle but they create some havoc while they bounce around during their brief lifespan. The churning quantum foam is believed to extend throughout the universe even filling the empty space within the atoms in human bodies.

1997 Lamoreaux verifies Casimir

Ground State of Hydrogen is Sustained by ZPE

1987 Puthoff finds electron-atomic energy state is ZPE

Looking at the electron in a set ground-state orbit, it consists of a bound state with a central Coulomb potential that has been treated successfully in physics with the harmonic oscillator model. However, the anomalous repulsive force balancing the attractive Coulomb potential remained a mystery until Puthoff published a ZPE-based description of the hydrogen ground state.[78] This derivation caused a stir among physicists because of the extent of influence that was now afforded to vacuum fluctuations. It appears from Puthoff's work that the ZPE shield of virtual particles surrounding the electron (Fig. 1.7) may be the repulsive force. Taking a simplistic argument for the rate at which the atom absorbs energy from the vacuum field and equating it to the radiated loss of energy from accelerated charges, the Bohr quantization condition for the ground state of a one-electon atom like hydrogen is obtained. "We now know that the vacuum field is in fact formally necessary for the stability of atoms in quantum theory."[79]

Lamb Shift Caused by ZPE

A milestone in quantum theory which led to a new appreciation for the fundamental nature of ZPE is called the "Lamb shift." Measured by Dr. Willis Lamb in the 1940's, it directly shows the effect of zero point fluctuations on certain electron levels of the hydrogen atom, causing <u>a fine splitting of the levels</u> on the order of 1000 MHz.[80] Los Alamos physicist Milonni says "The Lamb shift and its explanation marked the beginning of modern quantum electrodynamics."[81] It really showed how the atomic electron is coupled to the vacuum ZPF and is worthwhile reading about in any quantum physics book. Physicist Margaret Hawton describes the Lamb shift as "a kind of one atom Casimir Effect," which is profound. If you visualize the electron whirling around the nucleus with ZPE fluctuations that polarize the vacuum, the tendency is to push the electron into a slightly higher "orbit." Hawton agrees by stating that the vacuum fluctuations of ZPE need only occur in the vicinity of atoms or atomic particles for the Lamb shift to exist.[82] This idea also agrees with the discussion about Koltick in Chapter 1, illustrated in Figure 1.6.

1947 Lamb measures ZPE

1994 Hawton puts ZPE in atom

Today, "the majority of physicists attribute spontaneous emission and the Lamb shift entirely to vacuum fluctuations."[83] This may lead scientists to believe that it can no longer be called "spontaneous emission" but instead should properly be labeled forced or "stimulated emission" much like laser light, even though there is a random quality to it. However, it has been found that radiation reaction (the reaction of the electron to its own field) together with the vacuum fluctuations contribute equally to the phenomena of spontaneous emission.[84]

Experimental ZPE

The first journal publication to propose a Casimir machine for "the extracting of electrical energy from the vacuum by cohesion of charge foliated conductors" is shown in Figure 5.2 of this book.[85] Dr. Forward describes this "parking ramp" style corkscrew or spring as a ZPE battery that will tap electrical energy from the vacuum and allow charge to be stored. The spring tends to be compressed from the Casimir force but the like charge from the electrons stored will cause a repulsion force to balance the spring separation distance. It tends to compress upon dissipation and usage but expand physically with charge storage. He suggests using micro-fabricated sandwiches of ultrafine metal dielectric layers. Forward also points out that ZPE seems to have a definite potential as an energy source.

1984 Forward to a ZPE battery

Another intriguing experiment is the **"Casimir Effect at Macroscopic Distances"** which proposes observing the Casimir force at a distance of a few centimeters using confocal optical resonators within the sensitivity of laboratory instruments.[86] This experiment makes the microscopic Casimir effect observable and greatly enhanced. For those, like myself, who think ZPE energy conversion may be optimized at nanometer distances only, this discovery is really a head-turner. It means that with *creative vacuum engineering* the future energy Casimir motors are going to have parts *visible to the naked eye*.

In general, many of the experimental journal articles refer to vacuum effects in a cavity that is created with two or more surfaces. Cavity QED is a science unto itself. "Small cavities suppress atomic transitions; slightly larger ones, however, can enhance them. When the size of the cavity surrounding an excited atom is increased to the point where it matches the wavelength of the photon that the atom would naturally emit, vacuum-field fluctuations at that wavelength flood the cavity and become stronger than they would be in free space."[87] It is also possible to perform the opposite feat. "Pressing zero-point energy out of a spatial region can be used to temporarily increase the Casimir force."[88] The materials used for the cavity walls are also important. It is well-known that the attractive Casimir force is obtained from highly reflective surfaces. However, "...a repulsive Casimir force may be obtained by considering a cavity built with a dielectric and a magnetic plate. The product r of the two reflection amplitudes is indeed negative in this case, so that the force is repulsive."[89] **For parallel plates in general, a "magnetic field inhibits the Casimir effect."**[90]

1999 Pinto invents a ZPE engine

An example of an idealized system with two parallel semiconducting plates separated by an variable gap that utilizes several concepts referred to above is Dr. Pinto's **"optically controlled vacuum energy transducer."**[91] By optically pumping the cavity with a microlaser as the gap spacing is varied, the total work done by the Casimir force along a closed path that includes appropriate transformations is non-zero. This thermodynamic discovery means that major technological advances in the area of endless, free energy production can be achieved.[92] The barrier to utilizing ZPE has been broken. More analysis on this revolutionary invention is found in later chapters.

ZPE Patent Review

For any researcher reviewing the literature for an invention design such as energy transducers, it is vital to be thorough and perform a patent search. In 1987, Werner and Sven from Germany patented a "Device or method for generating a varying Casimir-analogous force and liberating usable energy" with patent #DE3541084. It subjects two plates in close proximity to a fluctuation which they believe will liberate energy from the zero-point field.

In 1996, Jarck Uwe from France patented a "Zero-point energy power plant" with PCT patent #WO9628882. It suggests that a coil and magnet will be moved by ZPE which then will flow through a hollow body generating induction through an energy whirlpool. It is not clear how such a macroscopic apparatus could resonate or respond to ZPE effectively.

On Dec. 31, 1996 the conversion of ZPE was patented for the first time in the United States with US patent #5,590,031. The inventor, **Dr. Frank Mead**, Director of the Air Force Research Laboratory, designed receivers to be spherical collectors of zero point radiation (see Figures 3.1 and 2.4). Since *it gets stronger the smaller you make it*, extremely high frequencies should be the best, the limit of which, by some estimates, corresponds to the Planck frequency of 10^{43} Hz. We do not have any apparatus to amplify or even oscillate at that frequency currently. For example, gigahertz radar is only 10^{10} Hz or so. Visible light is about 10^{14} Hertz and gamma rays reach into the 20th power, where the wavelength is smaller than the size of an atom. However, that's still a long way off from the 40th power. The essential innovation of the Mead patent is the "beat frequency" generation circuitry, which creates a lower frequency output signal from the ZPE input. When I called Frank and spoke to him about his patent, I happened to mention my idea that a proton and a neutron might seem to be the best combination to use, since they are slightly different in size and may be able to create the right kind of beat frequencies. However, his reaction was surprising because it seemed that I had guessed at

1996 Mead gets a ZPE patent

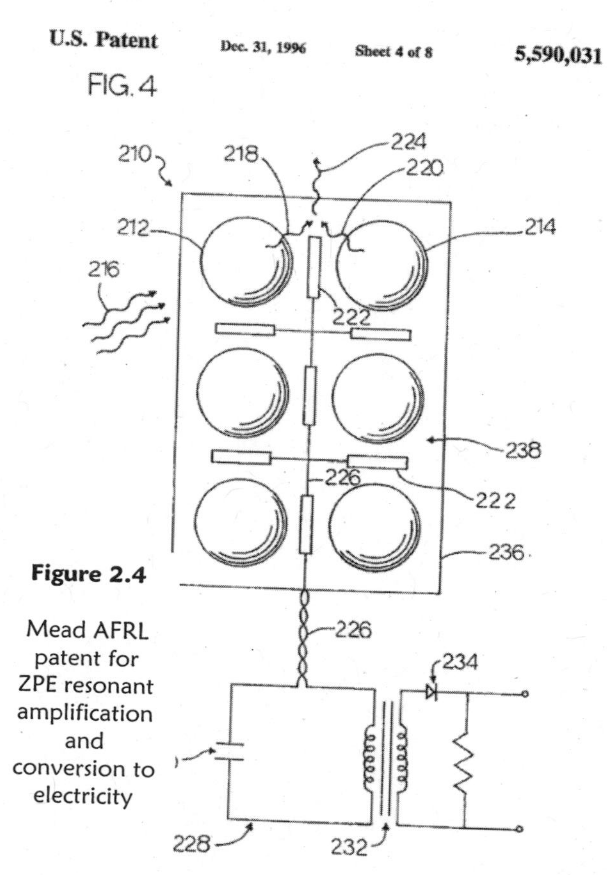

Figure 2.4

Mead AFRL patent for ZPE resonant amplification and conversion to electricity

something he regarded as a trade secret. The conclusions that I have drawn concerning Mead's patent are elaborated in my *Practical Conversion of ZPE* book.[93] To summarize, while it is an interesting concept, the beat frequency operation does not seem yield a significantly lowered frequency of resonance to warrant such a design, unless the two frequencies of the two spheres are so close that the subtraction of their resonant frequencies results in a thousand fold decrease in the beat frequency. It awaits a devoted nanotech engineer who is willing to bring a proton and neutron close together test the principle. Nature has already done the job by creating "deuterons" so instrumentation is all that is needed to perform the experiment!

Another patent that utilizes a noticeable ZPE effect is the AT&T "Negative Transconductance Device" by inventor, **Federico Capasso** (US #4,704,622). It is a resonant tunneling device with a one-dimensional quantum well or wire. The important energy consideration involves the zero-point energy that is available to the electrons in the extra dimensional quantized band, allowing them to tunnel through the barrier.

This solid state, multi-layer, field effect transistor *demonstrates that without ZPE, no tunneling would be possible.* It is supported by the virtual photon tunnel effect.[94]

Grigg's Hydrosonic Pump is another patent (U.S. #5,188,090), whose water glows blue when in cavitation mode, that consistently has been measuring an over-unity performance of excess heat energy output. It appears to be a dynamic Casimir effect that contributes to sonoluminescence.[95]

Joseph Yater patented his "Reversible Thermoelectric Converter with Power Conversion of Energy Fluctuations" (#4,004,210) in 1977 and also spent years defending it in the literature. In 1974, he published "Power conversion of energy fluctuations."[96] In 1979, he published an article on the "Relation of the second law of thermodynamics to the power conversion of energy fluctuations"[97] and also a rebuttal to comments on his first article.[98] It appears that energy is being brought from a lower temperature reservoir to a higher one, which normally violates the 2nd law. The basic concept is a simple rectification of thermal noise, which also can be found in the **Charles Brown** patent (#3,890,161) of 1975, "Diode array for rectifying thermal electrical noise."

Many companies are now very interested in such processes for powering nanomachines. While researching this ZPE book, I attended the American Association for the Advancement of Science (AAAS) workshop by IBM on nanotechnology in 2000. There I learned that R. D. Astumian proposed in 1997 to rectify thermal noise (as if this was a new idea).[99] **This had the direct effect of provoking IBM to begin a "nanorectifier" development program.**

In 2003, **Dr. Fabrizio Pinto** received two patents on his Casimir force transducer. The first one (#6,650,527) is very exciting because it explicitly identifies the two major ways that Pinto alters the Casimir force, by changing physical properties of the transducer, with tricks we will explore in later chapters. The abstract states:

2003 Pinto's ZPE Casimir patents

> Method and apparatus for modulating the direction and magnitude of the Casimir force between two bodies. In accordance with the illustrative embodiment, a repulsive Casimir force is generated by placing two bodies in near-proximity to one another. For one of the bodies, dielectric properties predominate; for the other, magnetic properties predominate. The arrangement further includes a device that alters the dielectric or magnetic properties of the bodies. By altering the dielectric or magnetic properties, the repulsive Casimir force can be made to decrease, then vanish, then reappear as an attractive force. Modulating the Casimir force in such a manner can be used to control stiction in MEMS devices and to accelerate particles, among many other applications.

The second patent (#6,665,167), granted a month later, has a very interesting abstract which simply patents the detailed method for energy extraction and does not define the ZPF but calls it a "force field."

> In some embodiments, the illustrative method defines an engine cycle comprising several state changes that allow for a net gain of energy from an underlying source force field. The potential for a net energy gain via the method results from the discovery that a Casimir force system can be rendered non-conservative. This is done by appropriately altering one or more of a variety of physical factors that affect the Casimir force, or by altering any of a variety of environmental factors that affect such physical factors. In various embodiments, the extracted energy is stored, used to power energy-consuming devices or used to actuate a micromechanical device.

This is also a milestone for the Casimir force being put to work and also creating the proper thermodynamic circumstances to ensure that ZPE is not conserved.

1996 Eberlein finds SL is ZPE

ZPE and Sonoluminescence

Sonoluminescence (SL) converts ZPE. This conclusion is based upon the experimental results of ultrasound cavitation in various fluids which emit light and extreme heat from bubbles

100 microns in diameter which implode violently creating temperatures of 5,500 degrees Celsius. Scientists at UCLA have recently measured the length of time that sonoluminescence flashes persist. Barber discovered that they only exist for 50 picoseconds (ps) or shorter (1 ps = 10^{-12} seconds = one millionth of a millionth of a second), which is too brief for the light to be produced by some atomic process. Atomic processes, in comparison, emit light for at least several tenths of a nanosecond (ns). "To the best of our resolution, which has only established upper bounds, **the light flash is less than 50 ps in duration** and it occurs within 0.5 ns of the minimum bubble radius. The SL flashwidth is thus 100 times shorter than the shortest (visible) lifetime of an excited state of a hydrogen atom."[100]

Critics of the obvious nature of this unique light spectrum however, still tried to propose other mechanisms, like atomic transitions to explain SL. **Dr. Claudia Eberlein** in her pioneering paper "Sonoluminescence and QED" describes her conclusion that only the ZPE spectrum matches the light emission spectrum of sonoluminescence, and could react as quickly as SL.[101] She thus concludes that SL must therefore be a ZPE phenomena. This breakthrough establishes one more unexpected ZPE link to the quantum vacuum.

Gravity is Related to ZPE

> **1989** Puthoff says gravity is ZPE

Another dimension of ZPE is found in the work of Dr. Harold Puthoff, who has found that gravity is a zero-point-fluctuation force, in a prestigious <u>Physical Review</u> article that has been largely uncontested.[102] He points out that the late Russian physicist, Dr. Sakharov regarded gravitation as not a fundamental interaction at all, but an induced effect that's brought about by changes in the vacuum when matter is present. The interesting part about this is that the mass is shown to correspond to the kinetic energy of the zero-point-induced internal particle jittering, while the force of gravity is comprised of the long ZPE wavelengths. This is in the same category as the low frequency, long-range forces that are now associated with Van der Waal's forces or "dispersion" forces.

Inertia is Caused by ZPE

Everyone has had the experience of going around a curve in a car a little too fast, as the car tries to fight the turn. This is what Newton called "inertia," the tendency of a body at rest to stay at rest but also for a body in motion to stay in a straight line direction of motion. The inertial force we feel pushing us and the car toward the edge of the road, causes car accidents on slippery highways. Referring to the inertial relationship to zero-point energy, Dr. Bernard Haisch at the Calphysics Institute finds that first of all, that inertia is directly related to the Lorentz Force which is normally used to describe Faraday's Law.[103] As a result of his work, the Lorentz Force now has theoretically been shown to be directly responsible for an electromagnetic resistance arising from a distortion of the zero-point field in an accelerated frame. He also explains how the magnetic component of the Lorentz force arises in ZPE, its matter interactions, and also a derivation of Newton's law, **F = ma**. From quantum electrodynamics, Newton's law appears to be related to the known distortion of the zero point spectrum in an accelerated reference frame.

1994 Haisch blames ZPE for inertia

Haisch et al. present an understanding as to why force and acceleration should be related, or even for that matter, what is mass.[104] Previously misunderstood, mass (gravitational or inertial) is apparently more electromagnetic than mechanical in nature. (The "equivalence principle" in physics says that gravitational and inertial mass are the same.) Now it is known that the resistance to acceleration defines the inertia of matter but interacts with the vacuum as an electromagnetic resistance. To summarize the inertia effect, it is connected to a distortion at high frequencies of the zero-point field. Whereas, the gravitational force has been shown to be a low frequency interaction with the zero point field. Read more at calphysics.org.

Dark Energy Can Be Measured in the Lab

2005 Beck says dark energy measured in lab

In 2005, Dr. Christian Beck, from King's College in London published an article proposing that published results from Koch prove that dark energy (which he said can be confirmed to be ZPE) has been measured in the lab.[105] I briefly corresponded with him after his article was published. (See Appendix for the abstract of his paper and commentary.)

While some may still question the availability of nonthermal fluctuations from the ZPF in a solid state device, Beck and Mackey experimentally measured the spectral density of current noise in Josephson junctions in 2004. They assert that it provides direct evidence for the existence of zero-point fluctuations. *Assuming that the total vacuum energy associated with these fluctuations cannot exceed the presently measured dark energy of the universe*, they predict an upper cutoff frequency of f_c = (1.69 +/- 0.05) x 10^{12} Hz for the measured frequency spectrum of zero-point fluctuations in the Josephson junction. This provides a reasonable resolution to one of the most hotly contested issues of ZPE: its cutoff frequency. Note that it is significantly less than the Planck cutoff frequency of f_c = 10^{43} Hz based on the Planck length discussed in Chapter 1.

Furthermore, Beck and Mackey help explain astronomy's self-created dilemma of dark energy that has remained unresolved because of misunderstandings of the properties of ZPE.[106] The largest frequencies that have been reached in the experiments are of the same order of magnitude as f_c and provide a lower bound on the dark energy density of the universe. They show that suppressed zero-point fluctuations above a given cutoff frequency can lead to 1/f noise. Therefore, it is quite conceivable that their experiment can measure some of the properties of dark energy in the lab.[107]

Lastly, it was quite entertaining for me to learn how Casimir also discovered the larger ZPE connections to his lab experiment from none other than **Neils Bohr** (the "Bohr model" of the atom is the introductory model taught in school). In 1999, Casimir reported the origins of his work.

> Here is what happened. During a visit to Copenhagen, it must have been in 1946 or 1947, [Neils] Bohr asked me what I had been doing and I explained our work on van der Waals forces. "That is nice," he said, "that is something new." I then explained I should like to find a simple and elegant derivation of my results. Bohr thought this over, then mumbled something like "must have something to do with zero-point energy." That was all, but in retrospect, I have to admit that I owe much to this remark.[108]

An article in *Physics Today* suggested years ago that **dark energy** and zero-point energy were the same. This was defended by a simple calculation that the author outlined to establish the similar properties and *sufficient energy density* for astronomical effects. It is likely that many astronomers will benefit from studying ZPE so that they can become familiar with the conditions necessary to **create the positive Casimir effect**, which, on a grand scale, is contributing to the repulsive force now seen accelerating the galaxies apart. Einstein simply used a "cosmological constant" to account for such an effect but that does not explain it. Even today, astronomers would rather leave the energy "dark" and mysterious instead of postulating as mathematicians and physicists have that dark energy is actually caused by ZPE. For all practical purposes, it is one and the same. I predict that this will be borne out by further experiments and education of astronomers in the near future.

NASA Supports ZPE

Besides a grant awarded to Jordan Maclay for Casimir research mentioned previously, NASA maintains a website that includes a description and short history of ZPE research.[109] Recently various conferences have lent support to the ZPE revolution by featuring lectures on the Casimir effect, energy development from the ZPF, and field propulsion proposals. One example is the Joint Propulsion Conference (JPC), which is an intersociety conference that also has Marc Millis from NASA's Breakthrough Propulsion Program chairing a session. In 2001, I met him as IRI collaborated with NASA to bring two Russian scientists to the conference to present their experimental results on a magnetic energy converter (MEC).[110] Though Marc told me that he is interested mostly in "incremental improvements" his efforts have largely helped to attract proposals for disruptive technologies as well, which are the more desirable ones for significant change in society. However, in 2006, Marc presented a paper at JPC on "Responding to Mechanical Antigravity" which debunked every inertial propulsion invention he could find with the words, "errant claim."[111] When asked by me, his reaction to the Russian presentation in 2001 was "too many outrageous claims" which now in retrospect seems like a habit of his.

Marc Millis at JPC in 2001

US005590031A

United States Patent [19]

Mead, Jr. et al.

[11] **Patent Number:** **5,590,031**

[45] **Date of Patent:** **Dec. 31, 1996**

[54] **SYSTEM FOR CONVERTING ELECTROMAGNETIC RADIATION ENERGY TO ELECTRICAL ENERGY**

[76] Inventors: **Franklin B. Mead, Jr.**, 44536 Avenida Del Sol, Lancaster, Calif. 93535; **Jack Nachamkin**, 12314 Teri Dr., Poway, Calif. 92064

[21] Appl. No.: **281,271**

[22] Filed: **Jul. 27, 1994**

[51] **Int. Cl.**^{6} **H02M 1/00**

[52] **U.S. Cl.** **363/8**; 363/178; 342/6

[58] **Field of Search** 363/8, 13, 178; 342/6, 61, 73, 173, 175

[56] **References Cited**

U.S. PATENT DOCUMENTS

3,882,503	5/1975	Gamara	343/100 R
4,725,847	2/1988	Poirier	343/840
5,008,677	4/1991	Trigon et al.	342/17

Primary Examiner—Peter S. Wong
Assistant Examiner—Adolf Berhane
Attorney, Agent, or Firm—Chris Papageorge

[57] **ABSTRACT**

A system is disclosed for converting high frequency zero point electromagnetic radiation energy to electrical energy. The system includes a pair of dielectric structures which are positioned proximal to each other and which receive incident zero point electromagnetic radiation. The volumetric sizes of the structures are selected so that they resonate at a frequency of the incident radiation. The volumetric sizes of the structures are also slightly different so that the secondary radiation emitted therefrom at resonance interfere with each other producing a beat frequency radiation which is at a much lower frequency than that of the incident radiation and which is amenable to conversion to electrical energy. An antenna receives the beat frequency radiation. The beat frequency radiation from the antenna is transmitted to a converter via a conductor or waveguide and converted to electrical energy having a desired voltage and waveform.

14 Claims, 8 Drawing Sheets

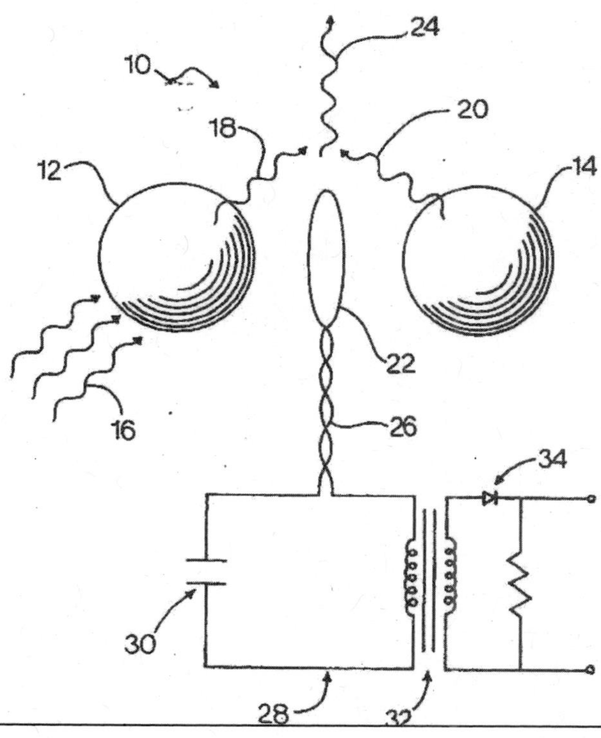

Figure 3.1 Dr. Frank Mead's ZPE patent cover page.

Chapter 3

ZPE as an All-Pervading Electromagnetic Field

Overview

Years ago, when I was a senior physics undergraduate at the State University of NY at Buffalo, Professor Y. C. Lee introduced zero-point energy in a course in Quantum Mechanics. It was the first and only time that God was ever mentioned in any physics class, as Dr. Lee drew a comparison to the **"all-pervading electromagnetic field"** of zero-point energy. I was astonished but appreciated his spiritual aspiration for a topic which instantly became a subject of intrigue for me (see Ch. 8 for more on this). While the theoretical basis for the electromagnetic nature of ZPF is rigorous, even in classical physics, the concept of such an electromagnetic field, like light, pervading everything in the universe, may be comforting or confusing to the reader. For me it is comforting and the more I learn about it, the more interesting it seems to be. In other words, with a minimum of quantum physics, the ZPF can be regarded scientifically as simply electromagnetic energy. This is the insight that Dr. Frank Mead exploited for his ZPE patent (Figs. 2.4 and 3.1), which is discussed below.

Electromagnetic Energy is Everywhere

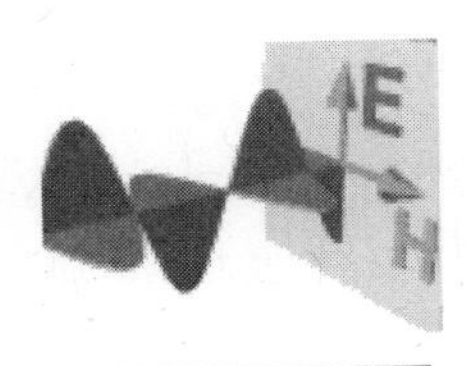

Fig. 3.2
EMF Wave

Everything in nature responds to an electromagnetic field (EMF) and exchanges EMF energy all day long. Figure 3.2 shows the standard model for two waves of electric and magnetic oscillation as they travel at the speed of light. Our cells absorb it from sunlight and store the energy in their DNA. Another book of mine explores bioelectromagnetics and the beneficial EMF exchange in living beings called "biophotons."[112] Plants that convert it into sugar for storage are great at taking advantage of its energy content. Another aspect of the sunlight's EMF involves the reason that the grass is green. The plant cells have just absorbed mostly ultraviolet (UV) EMF energy from the sunshine, keep some of it and then re-radiate it back at a lower frequency energy in the visible spectrum (see Fig. 3.3) which is interpreted by our brain as "green." Another example of EMF energy is when people are gathered in a room or home, environmental experts use a rule of thumb that each person's radiated energy is equivalent to an old 20th century, 100 Watt incandescent light bulb. In other words, every person's body on the average, radiates as much thermal infrared energy (IR) as the light bulb, which will tend to heat up the room or home. (In

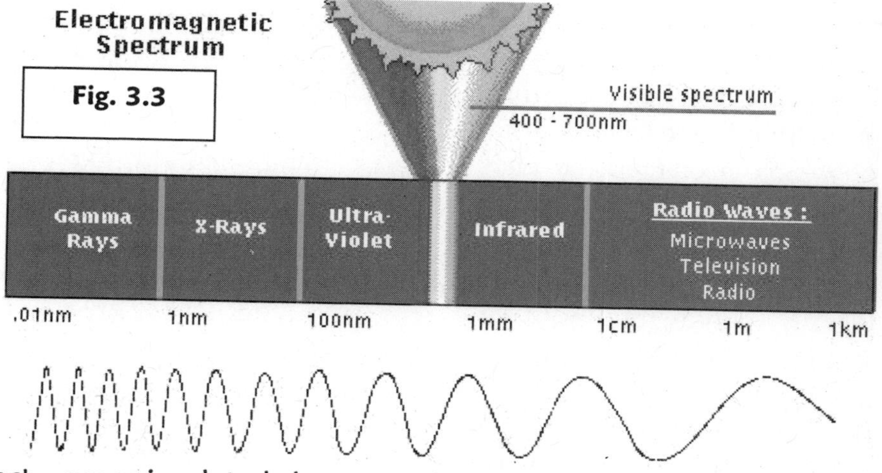

fact, super-insulated homes take advantage of this fact by trapping that heat and not letting it escape, thus saving on heating bills.)

All of the EMF energy mentioned thus far however, is manufactured from electrons in upper energy levels of an atom as they drop down to lower levels and re-radiate that energy difference, in a matter of **microseconds**. This is also where chemical energy exchange originates from and has very little to do with ZPE. Some may note that the EMF photons are using the same medium to transmit the EMF wave or photon, which is like a wave traveling on the ocean. (In Chapter 1, we mentioned the "aether" or "ether" as one theoretical model for the ZPE medium we normally call the "quantum vacuum.") If something different than electron levels were responsible for the EMF transmission, one clue might be that the time delay between the stimulation and the emission was very short...much shorter than any possible electron transition time. This experimental evidence has already occurred in the laboratory. Quite different that even high frequency gamma rays that are emitted from radioactive atoms, short **picosecond** bursts of light have been seen in experiments involving "sonoluminescence."

Further Details on ZPE and Sonoluminescence

The concept that sonoluminescence (SL) taps ZPE was proven in the last chapter. The SL electromagnetic radiation flashwidth is 100 times shorter than the shortest (visible) lifetime of an excited state of a hydrogen atom."[113]

Critical to the understanding of the nature of this light spectrum however, is what other mechanism than atomic transitions can explain SL. It is also acknowledged that "Schwinger proposed a physical mechanism for sonoluminescence in terms of photon production due to changes in the properties of the quantum-electrodynamic (QED) vacuum arising from a collapsing dielectric bubble."[114] As mentioned in the previous chapter, Dr. Claudia Eberlein in her pioneering paper "Sonoluminescence and QED" describes her conclusion that **only the ZPE spectrum matches the light emission spectrum of sonoluminescence**, *and could react as quickly as SL*. She is forced to conclude that SL must therefore be a ZPE phenomena causing electromagnetic light production due to changes in the properties of the

Dielectric - An electrically-insulating substance often used in capacitors to help store more charge. Glass or water are dielectrics which can absorb, transmit, or reflect EMFs quantum-electrodynamic (QED) vacuum arising from a collapsing dielectric bubble.[115] Therefore, the ZPF can produce light under the right circumstances, which also is an energy release.

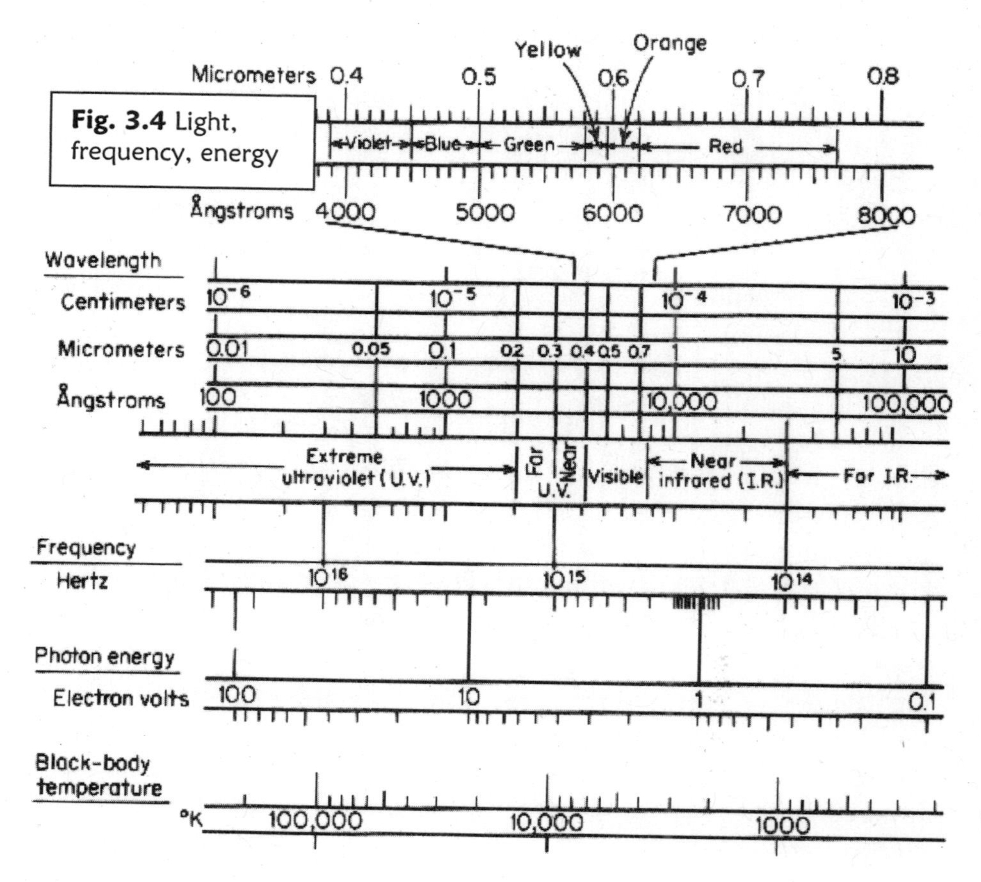

Fig. 3.4 Light, frequency, energy

Air Force Research Lab Creates ZPE Converter

Treating the quantum vacuum initially as an all-pervading electromagnetic wave with a high bandwidth is a classical physics approach. In other words, no modern physics is required to understand a bunch of EMF waves coming in from everywhere. It's a no-brainer. This is not implying anything about the US Air Force but this approach certainly simplifies things quite a bit. The first invention to use this method for ZPE is a U.S. patent (#5,590,031) proposing **microscopic antennae** for collecting and amplifying zero-point electromagnetic energy. It is a US Air Force invention by Mead et al. that offers sufficient scientific rigor and intrigue to warrant an investigation by the reader (Note: patent copies are available for free at Google.com/patents or www.uspto.gov). The patent's spherical resonators are small scatterers of the zero-point vacuum flux and capitalize on the electromagnetic wave nature of the ZPF. Utilizing this design can start an inquiry at least into the microscopic and nanotechnology realm. (Detailed analysis of this patent is available in my other ZPE book, *Practical Conversion of Zero Point Energy*.[116])

The Mead patent basically uses dielectric spheres to scatter ZPE EMFs and convert them to electricity. I feel that the general design criteria of the patent is feasible. However, scattering of the EMFs can reduce the benefit of the resonant amplification. Interestingly, **smaller spheres are more desirable because they are more energetic**[117] The smaller diameter means that shorter wavelengths and higher frequencies are resonant with them. Impressed with the simplicity of the model, I became intrigued with the invention and wondered which size of dielectric sphere would be the best. Separating them into four spheres of interest, I came up with the following categories:

1) **microsphere**: micron-sized (10^{-6} m) electrolithography;
2) **nanosphere**: nanometer-sized (10^{-9} m) molecular nanotech;
3) **picosphere**: picometer-sized (10^{-12} m) atomic technology; and
4) **femtosphere**: femtometer-sized (10^{-15} m) nuclear technology.

It can be concluded that the general design criteria of the patent is feasible, as Dr. Mead states:

> "the spheres must be small in direct proportion to the wavelength of the high frequencies of the incident electromagnetic radiation at which resonance is desirably obtained" (col. 7, line 66).

However, as we try to take advantage of the "third-power" increase in energy density available from ZPE by decreasing the size, there comes a point where the best performance and feasibility comes from a proton stuck to a neutron (called a "deuteron"). This is the smallest and yet ironically, *the most energetic receiver available* to ZPE. The reason is that as the spectral energy density increases, it depends upon the third power (cube) of the frequency. Therefore, it is very attractive in all ZPE designs to **make the converter as small as possible**. As an aside, is it any wonder that cold fusion experiments have used deuterons and deuterium almost invariably? It may be possible that ZPE contributes to the energy generation, which the numerous government labs report.[118]

One of the interesting considerations was to design the Mead receivers for the range of extremely high frequency that ZPE offers, which by some estimates, corresponds to the Planck frequency of 10^{43} Hz. We do not have any apparatus to amplify or even oscillate at that frequency currently. For example, gigahertz radar is only 10^{10} Hz or so. Visible light is about 10^{14} Hertz and gamma rays reach into the 20th power, where the wavelength is smaller than the size of an atom. However, that's still a long way off from the 40th power. The essential innovation of the Mead patent is the "beat frequency" generation circuitry (the beat goes on☺), which creates a lower frequency output signal from the subtraction of the two ZPE input signals (creating a "difference" signal).

Focus EMFs from ZPE

Reviewed in Chapter 9 in detail, focusing vacuum fluctuations is the same type of operation that is normally performed on EMFs. Regarding ZPE as electromagnetic waves allows such surfaces to actually reflect and focus the virtual EMF energy to a point. This type of parabolic mirror, designed microscopically as many microwave diode receivers are, can intensify the amount of energy reaching the receiver. When we consider the various types of rectifying diodes that can be used for converting thermal and non-thermal noise from ZPE to direct current (DC), it is good to keep in mind that parabolic mirrors will improve any conversion design, even if both are very microscopic or nanoscopic, as long as the reflecting surface has a high reflectivity in the highest frequency ranges.

Summary

This serves as a brief introduction to the simple but powerful concept of treating ZPE as EMFs. Hopefully the reader will be more comfortable with the subject with such a familiar basis for its existence. Years ago, when I told Hal Puthoff that the Mead patent was the first ZPE patent, he replied, "I have the first one." Upon researching the patent literature, the #5,208,844 or #5,018,180 patents for "discrete, contained charged particle bundles" from 1991 and 1993 are apparently the ones he was referring to. These co-pending patents describe the charge clusters as bundles of electrons that can be converted to energy by an X-ray emitting target or simply by conversion to heat. Though the vacuum ZPF does not seem to be mentioned in the patents, it is arguable that it is involved, since Ken Shoulders, who also patented the same phenomenon, has recently been co-authoring an article about the reliance of charge clusters or "exotic vacuum objects" on the ZPF. Further information on this exciting energy source is available from Ken's website http://svn.net/krscfs and also from Dr. Hans Dehmelt, who in 1995 announced in a journal that he had produced "kilo-electron balls."[119]

It is also worth mentioning an implied endorsement for ZPE research that comes from the US government, which also likes this approach. A few years ago, the USAF paid Dr. Forward to produce a report with a description of ZPE in terms of EMF frequency modes.[120] The AFRL also paid Dr. Mead to patent the EMF patent for converting ZPE and NASA gave a $300K grant to Dr. Maclay for ZPE research. As the US government moves ahead with this research, it is important for the civilian sector to also benefit from new energy discoveries.

Zero Point Energy

(Emerging science, 1948...)

What?

- Random Electromagnetic waves remain after all energy is removed
- Enormous energy density: 10^{24} to 10^{58} Joules/m^3
- Theorized to indirectly cause gravity and inertia

Why?

- As an energy source?
- As a reactive medium?

Evidence?

- Casimir Effect
- Plank blackbody spectrum
- quantum effects

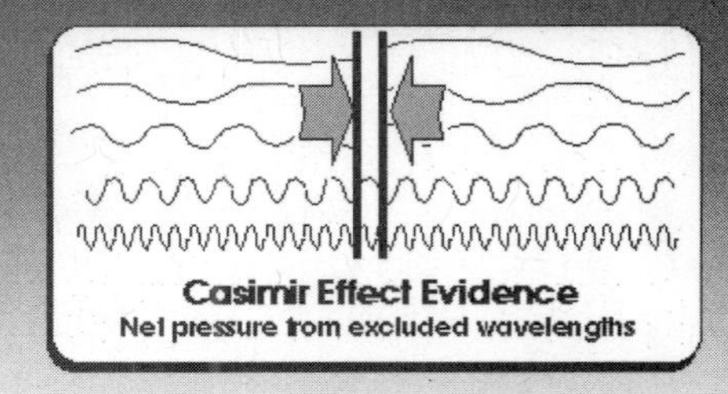

Casimir Effect Evidence
Net pressure from excluded wavelengths

Figure 4.1 Random EMF noise is characteristic of zero point energy according to this NASA ZPE web page, which is part of their *Breakthrough Propulsion Program*: www.grc.nasa.gov

Chapter 4

What's All of that Noise and What Good is it?

Overview

Everyone over a certain age knows what noise is. Parents complain to kids about it and neighbors cite the town's "noise ordinance" to fight the stuff. However, in electronics, it is more specifically that annoying random static on any radio of television receiver that uses an antenna (Fig. 4.2). If you tune your MP3 player to an FM station that is not coming in very well, you will discover noise. Did you know that noise comes in colors? Yes, it does. The best-known example is "white noise" which is a hissing sound that almost everyone has heard on radio receivers. This white noise tends to be evenly distributed across most of the measurable frequencies, like white light. So it is also called white. Another lesser-known example is "pink noise" which is more heavily weighted toward the lower frequencies. But where does it come from?

Fig. 4.2 Johnson noise on a television due to heat and ZPE

Another clue comes from the description of the noise in question, which is so fundamental that is persists even when thermal noise (called "Johnson" noise) is eliminated. This noise tends to be the highest in amplitude at the lowest frequencies. Furthermore, as we go up in frequency the noise reduces in proportion to 1/f where f is the frequency (when plotted on a log-log graph as in Fig. 4.3[121] – remember "logarithms" from high school?).

One of the explanations from the semiconductor industry for this noise is "electron relocation." This can be thought of as conduction electrons bouncing around trying to find a secure place to sit. What causes them to bounce around? Furthermore, what if we make it really cold so no Johnson noise is contributing to the 1/f noise? These are important questions to consider but many of us simply just want to know what good is it?

If we maintain the mystery a while longer (most engineers and scientists have done so for all of their lives), you will see how the picture starts to clear up to include ZPE.

Noisy Stuff

In electronic devices, the noisy stuff comes in even more delicious flavors. The typical kinds are shot noise, telegraph noise, and the ones mentioned above, Johnson noise that often

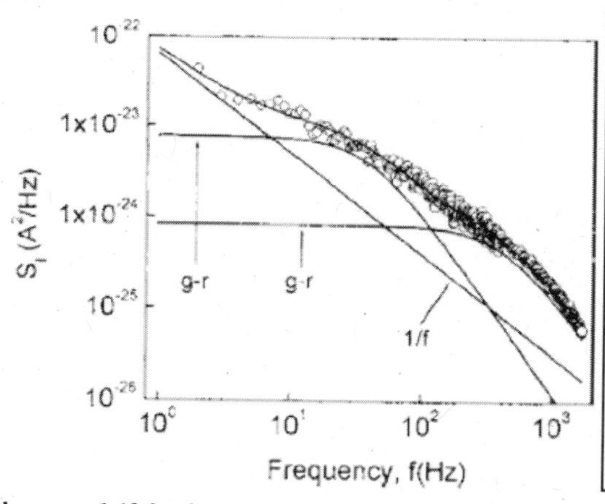

Figure 4.3

A good example of the experimental decrease of noise with increasing frequency in diodes. Note the ideal "1/f" noise predicted with the straight line. The total noise is a composite of two quantum dots (g-r lines) plus the diode's 1/f noise.

has a 1/f behavior. What is all of that noise? Random telegraph noise for example comes from an electron trapped in a quantum dot that can switch between two discrete states, so it sounds like a telegraph "dot, dash" if put on a loud speaker. Shot noise is when the drift velocity of electrons is subjected to an electric field. However, the most important question for our purposes is where does it come from and *can it be put to work?* A recent article and patent declare that the origin of 1/f noise, also known as "flicker noise," has remained "elusive" which is surprising for engineers to admit,[122] with an origin that is not precisely known.[123] However, there are some juicy clues to its nature to consider.

If You Dissipate, You Must Fluctuate

First, let's look at my favorite theorem that forms the foundation of ZPE in physics, which is reviewed in detail in my *Practical Conversion of ZPE* book. To briefly acquaint you with this eye-opener, it is called the "**fluctuation-dissipation theorem**." It is actually quite simple and widely applicable in all sorts of systems. This excites me when such a *system science principle* is explained and proven to be true because it feels like I am peering behind the veil of the material world into its underpinnings. The theorem establishes an exact mathematical relationship between the "impedance" or resistance in a system that dissipates (like a leaky bucket) and the fluctuations (remember quantum or thermal fluctuations?) of appropriate generalized "forces." The theorem itself is expressed as a single equation, essentially the same form as the original formula by Johnson[124] from Bell Telephone Laboratory who in 1928 discovered the thermal agitation "noise" of electricity.

Now that we got through that explanation, you'll be happy to learn that the theorem only has three variables. However, I am reminded to throw in some humor that is actually relevant, to get you started. Let's think of a two variable system first: *the immovable object gets hit by the irresistible force* ☺. This system doesn't change too much since the two variables are equally matched and very powerful. Our Johnson system of fluctuation and dissipation has three *correlated variables*,

1) **V** is the average (RMS) **value** of the spontaneously fluctuating force (which also could be **voltage** or electric field of virtual electrons),

2) **R** is the generalized impedance (**resistance**) and irreversible dissipation of the system,

3) **E(T)** is the average **energy** at temperature T of an oscillator of natural frequency.

The important conclusion to this rather long explanation is that all three are intimately correlated. If one changes, the other two have to change. If the force is identified in any system and it fluctuates, the resistance will have to be a dissipative loss of energy. In fact, Callen and Welton who discovered this theorem stated in their conclusion that this theorem also applies to the quantum vacuum. They said it also means that Planck has something to do with the average energy E(T):

> The existence of a radiation impedance for the electromagnetic radiation from an oscillating charge is shown to imply a fluctuating electric field in the vacuum, and application of the general theorem yields the Planck radiation law.[125]

This is the profound conclusion from their equation that the average energy E(T) for ZPE is equal to a summation (integral) over all the available frequencies of V times R. Their language of imperatives shows how inexorable this three-variable theorem really is, and therefore, the fundamental nature of the ZPF.

To illustrate, an example of the same type of relationship exists between the **three major variables in the earth's climate** which also are intimately correlated: 1) earth's average atmospheric temperature, 2) earth's sea level, and 3) the earth's CO_2 level. Don't believe it? Take a look at a composite diagram of the Vostok ice core measurements and how the three have tracked each other – inextricably linked – for the past 400,000 years (Fig. 4.3). Notice how the three lines keep tracking each other closely over a wide range of changes. Of course this has very important implications for the earth's sea level and the earth's temperature, which will inevitably catch up to the present CO_2 level (unless we bring it back down soon). That's why Al Gore and NASA's climatologist Jim Hansen are very concerned about this graph (see www.climatechange.org for more info).

Therefore, it should be much more understandable how a value of a dissipative **force**, a generalized **resistance**, and an average **energy** could be intimately linked in a quantum dance together. The actual dissipative force for ZPE is "radiation reaction" which is more fully explained in Milonni's book (see below) and also my *Practical Conversion of ZPE* book.

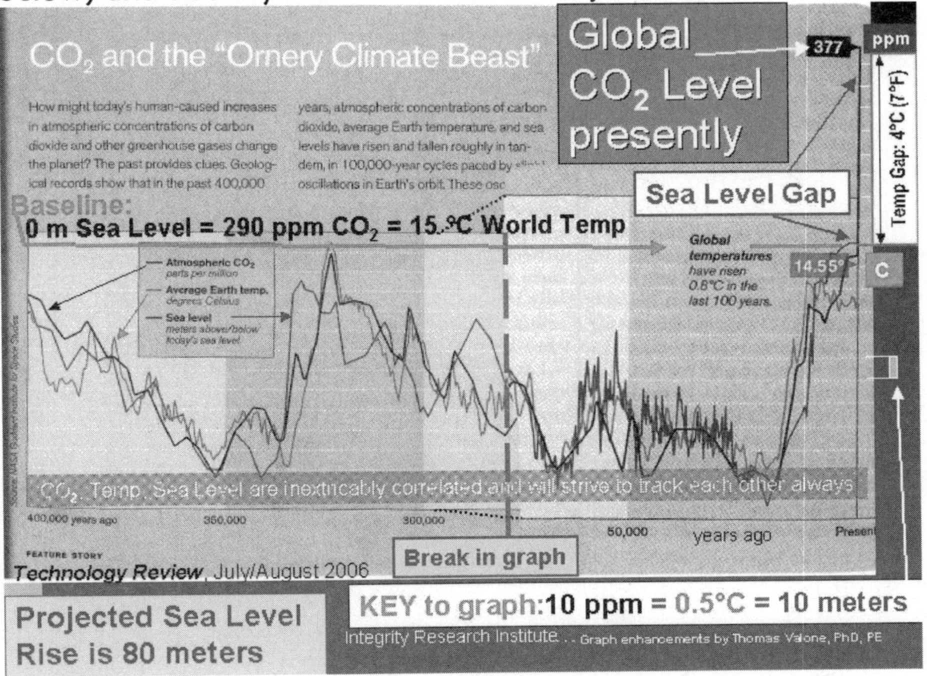

Fig. 4.3 Record of earth's last 400,000 years, from *Technology Review* magazine, compressed and annotated by the author, showing a different system with a **three-variable correlation** that cannot be separated. (Image can be downloaded for free from publisher IRI's website)

The famous physicst, Dr. Peter Milonni, from Los Alamos Laboratory, wrote the best textbook on ZPE, called *The Quantum Vacuum*.[126] His book was instrumental in my decision to pursue a PhD thesis on ZPE and to write this book to make the information more accessible to the layperson. What does this world-class expert on ZPE say about the existence of the

ZPE vacuum in connection to the **fluctuation-dissipation theorem**?

> "Generally speaking, if a system is coupled to a 'bath' that can take energy from the system in an effectively irreversible way, then the bath must also cause fluctuations. The fluctuations and the dissipation go hand in hand; we cannot have one without the other...*the coupling of a dipole oscillator to the electromagnetic field has a dissipative component, in the form of radiation reaction, and a fluctuation component, in the form of zero-point (vacuum) field;* given the existence of radiation reaction, the vacuum field must also exist in order to preserve the canonical commutation rule and all it entails."[127]

Though the language is quite formal, the implications are clear: *the quantum vacuum must fluctuate.* If you remember nothing else from this book, the fluctuations of ZPE should be indelibly imprinted on your mind. To me, this is the key to an energy solution. Just like a refrigerator or heat pump forces heat to go in one direction only, this "rectification" of heat can also be applied to non-thermal fluctuations (see next chapter).

Before we move on to the vacuum engineer's hardware needed to do the work, there is one more thought about **amplifying spontaneous emissions (ASE)** that is quite helpful in this work. Amplifying the available ZPE fluctuations in a cavity (which can be a box or container) is noted in Milonni's text, "the vacuum field may be amplified...if the spontaneously emitted radiation inside the cavity is amplified by the gain medium, then so to must the vacuum field entering the cavity. Another way to say this is that **'quantum noise' may be amplified.**"[128] This will be an important vacuum engineer's tool to call upon when a circuit component, like a diode, is chosen to do the work of converting ZPE into electricity.

Thermal noise voltage

$V^2 = 4k_BTRB$

R is the device resistance,
B is the bandwidth in Hertz,
T is the absolute temperature in Kelvin, and
k_B is a constant called the Boltzmann constant.

Noise in Coils

When Johnson did his pioneering research to discover the fundamental noise of the universe, he worked with standard circuit components like an induction coil. Recently Dr. Eric Davis from the Institute for Advanced Studies in Austin has reported at the Space Technologies Application International Forum (STAIF) that their laboratory research with coils will pursue measuring quantum noise associated with ZPE, based on support from Lockheed Martin.[129] Based on a proposal by Blanco et al., theory predicts that multiple-turn coils of large radius and low resistance might amplify ZPF-induced voltage fluctuations by about 100 times.[130]

ZPE Noise in Rectifying Diodes

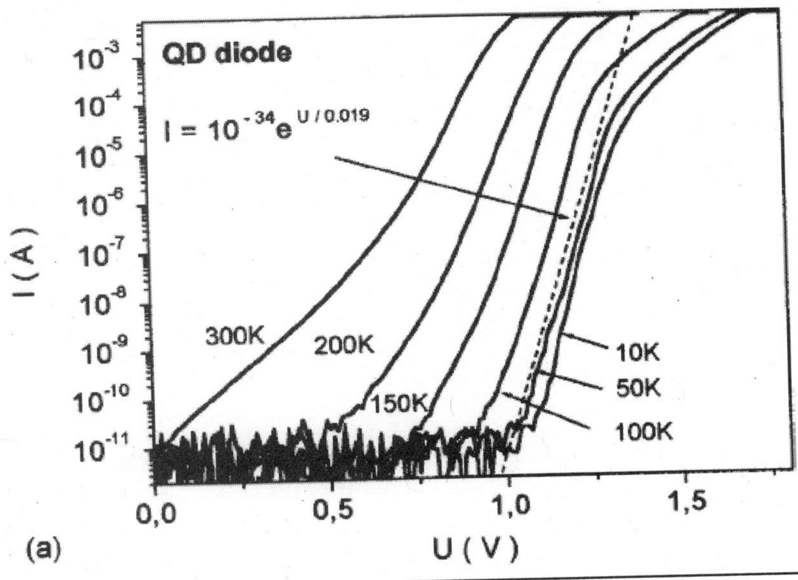

Fig. 4.4
A quantum dot diode with excess tunneling that has noise current even with no bias voltage, which does not depend on temperature.

Dr. Davis relies upon the published work of Koch, who did experiments with Josephson junctions (superconducting). They displayed quantum noise that was measurable as the virtual electrons tunneled through the insulating barrier of the Josephson junction.[131] Koch et al. also measured the frequency spectrum of the current fluctuations. Interestingly, "at low temperatures and high frequencies, **the experimental spectrum is dominated by zero-point fluctuations**."[132] This is a big discovery. It means that ZPE is confirmed in solid-state devices. Prof. Christian Beck, author of *Spatio-Temporal Chaos and Vacuum Fluctuations of Quantized Fields*, regarding Koch, states, "Zero-point fluctuations thus have theoretically predicted and experimentally measured effects in Josephson junctions."[133] He also emphasizes that zero-point energy has been important in predicting the spectrum of noise in many electrical circuits and also is the same **dark energy** present throughout the universe. Michael Turner from Fermilab agrees that dark energy is directly related to ZPE because "virtual pairs that fill up the vacuum have negative pressure" and, after a little calculation, "quantum vacuum energy is very repulsive."[134] He also shows how, ZPE pressure vs. energy density will change with an expanding universe, meeting an important requirement for dark energy. (See Beck's and Koch's reprinted articles on p. 198)

In Fig. 4.4 we see an example of a quantum dot diode that also exhibits noise, even when the voltage is at zero (note the jiggly lines at the bottom of the graph). These are subthreshold fluctuations at the level of 10 pA that can be attributed to quantum ZPE.[135] This appearance of thermal and non-thermal noise at zero bias has been seen in many other diodes as well, such as metal-oxide-metal (MOM) diodes.[136] The authors state that the noise changes to shot noise if a voltage bias is applied, thus demonstrating the connection between zero-bias conditions and thermal noise.

To appreciate a broader sense of the argument that is being presented here, there is a subject known as "**stochastic resonance**." It is a science that studies how the addition of noise to a non-linear system can improve its ability to cross a potential barrier and even behave in a resonant fashion. "The barrier-crossing rate thus depends critically on the noise intensity."[137] This phenomenon is directly analogous to the behavior of electrons in a zero-bias diode. With the addition of quantum noise and thermal noise, the electrons are able to cross the diode junction in a random (stochastic) fashion, based on the level of noise. In Fig. 4.5, we see and example of a "stochastic ratchet" which is just a noise-driven one-particle system that achieves movement in one direction (rectification).[138]

Summary

By digesting the complex topic of noise and diodes, the reader now has a taste of the parameters of likely technology that will encompass the future solid-state ZPE converter. Armed with this education, you have passed a milestone: <u>you are now officially a certified *Junior Vacuum Engineer*</u> and are cleared to proceed to the next level of advancement with the future energy training in the following chapters ☺.

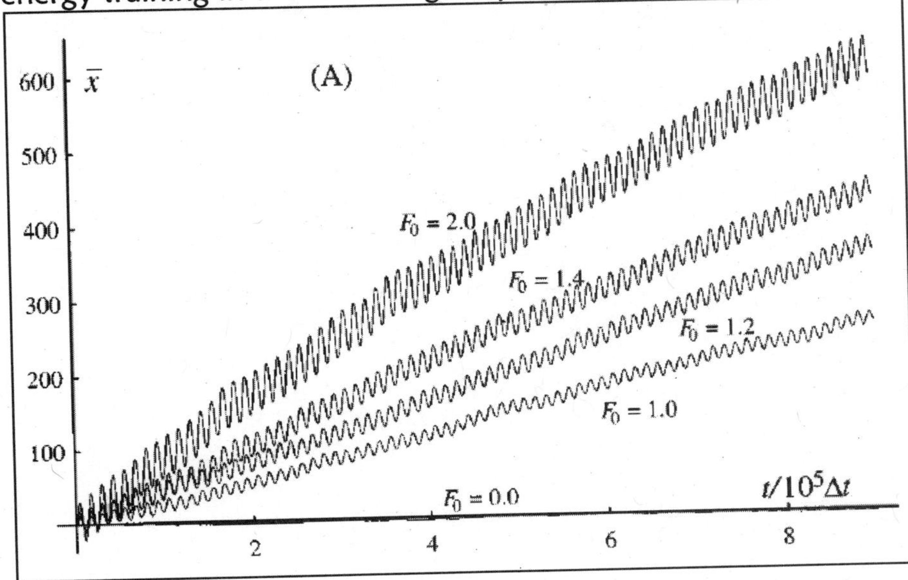

Fig. 4.5

Movement of a particle in the x direction versus time t of a thermal noise driven ratchet, with different external forces F_o

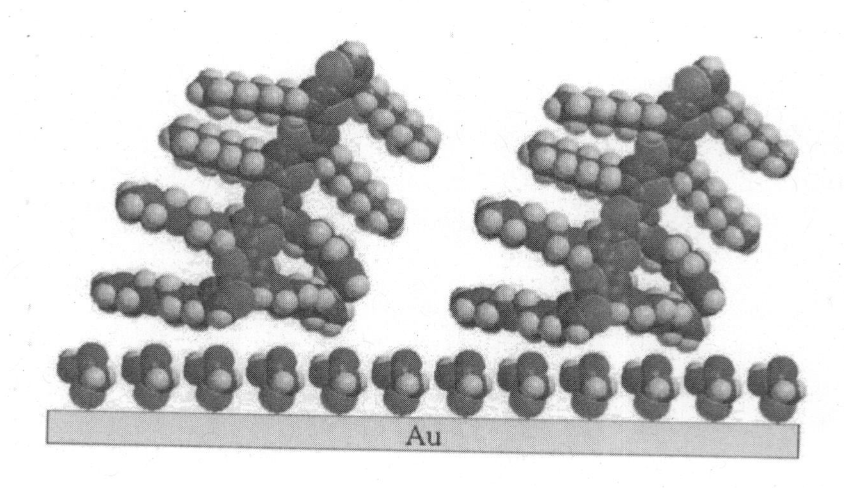

Figure 5.1 Dr. Yu's self-assembled molecular diode rectifiers on a gold substrate. The molecules line up all by themselves in this orientation so the contact to the top of each of them is easy (see Fig. 5.6).

Chapter 5

Zero Point Energy Can Supply Electricity

Motivation

Ever since the Second World War, we have had a problem with motivating physicists to band together and get back into a Manhattan Project of lasting consequence for our energy needs. Most of them, even some of the most brilliant minds in the field, would rather theorize about strings, axions, p-branes and quarks but not about new energy sources and new conversion methods. Credit must be given to a few in the academic and government labs who have made significant strides toward devices like molecular nanotechnology motors and cantilevers or zero-bias diodes that one day, may be mass produced into arrays, just like the tiny LEDs in flat screen TVs. However, most of the public still doesn't recognize that we have an energy crisis on our hands that is manifesting as climate change and severe global warming.

One of the most amazing insights that can be gained as we research the strange and wonderful world of ZPE conversion is that sometimes physicists and engineers overlook the obvious, perhaps since they are not motivated with grants, awards, or salary bonuses to solve the problem. For example, some of them just stop thinking with a comment like, "I have no idea how ZPE can be converted."

Below is a good example, still posted on a Department of Energy website, of the negative thinking that simply stops "experts" from being creative and pro-active because of their preconceived bias. Leave it to precocious 15-year-olds to ask the most embarrassing and highly intelligent questions!

"Ask A Scientist©"

Zero Point Energy

>>> name LB
> > status other
> > age 15

> > **Question - Would Zero-Point Energy be a better source of power than Antimatter?**
>
>Zero-point energy, the kinetic energy of a particle in its lowest quantum state, is a lousy source of energy. In fact, it's no source of energy at all. Since the object is already in its lowest quantum state, it's impossible to make it give up any of its energy. It would have no states of lower energy available to it!
>
>The only way would be to "open up" the particle's potential function, to give it new states of lower potential energy. I don't know if anyone has any ideas of how to do this; it might be possible, but I can't imagine how it could be practical.
>
>I can't imagine how antimatter could be practical, either. But perhaps sometime in the future, a source of antimatter can be found, as well as a means to contain it until it is used for generating energy. Certainly, it's well beyond current technical capabilities.
>
>Richard E. Barrans Jr., Ph.D.
>Assistant Director
>PG Research Foundation, Darien, Illinois
==
==================
>No, although neither seem promising at present.
>
>Tim Mooney
==
==================

>No. There would be no way to extract the zero point energy from a molecule in order to use it.

>Dr. Bradburn

Department of Energy
http://www.newton.dep.anl.gov/askasci/phy00/phy00034.htm

How to Convert Zero Point Energy

The first journal publication to propose a Casimir machine for "the extracting of electrical energy from the vacuum by cohesion of charge foliated conductors" is summarized here. Dr. Forward describes this "**parking ramp**" style corkscrew or spring as a ZPE battery that will tap electrical energy from the vacuum and allow charge to be stored (Fig. 5.2). The spring tends to be compressed from the Casimir force but the like charge from the electrons stored will cause a repulsion force to balance the spring separation distance. It tends to compress upon dissipation and usage but expand physically with charge storage. He suggests using micro-fabricated sandwiches of ultrafine metal dielectric layers. Forward also points out that ZPE seems to have a definite potential as an energy source.

Another interesting experiment is the "Casimir Effect at Macroscopic Distances" which proposes observing the Casimir force at a distance of a few centimeters **using confocal optical resonators** within the sensitivity of laboratory instruments.[139] This experiment makes the microscopic Casimir effect observable and greatly enhanced. Then, any other modes of conversion for the Casimir force should very well be improved.

In general, many of the experimental journal articles refer to vacuum effects on a cavity that is created with two or more surfaces. Cavity QED is a science unto itself. "Small cavities suppress atomic transitions; slightly larger ones, however, can enhance them. When the size of the cavity surrounding an excited atom is increased to the point where it matches the wavelength of the photon that the atom would naturally emit, vacuum-field fluctuations at that wavelength flood the cavity and become stronger than they would be in free space."[140] It is also possible to perform the opposite feat. "**Pressing zero-point energy out of a spatial region** can be used to temporarily *increase the Casimir force*."[141] The materials used for the cavity walls are also important. It is well-known that the attractive Casimir force is obtained from highly reflective surfaces. However, "...a repulsive Casimir force may be obtained by considering a cavity built with a dielectric and a magnetic plate. The product r of the two reflection amplitudes is indeed negative in this case, so that the force is repulsive."[142] For parallel plates in general, a "**magnetic field inhibits the Casimir effect**."[143] (See Appendix for the magnetic field connection to ZPE.) Just these tools alone are probably sufficient to build a simple but effective ZPE transducer. For example, if Forward's corkscrew ZPE battery was oscillated periodically within a cavity where the ZPE was pressed out and the Casimir force was increased during one half of the cycle, the spring would tend to attract and force electrons out of the battery by electrostatic repulsion. Then, during the second half of the cycle, electrons would be drawn back into the battery as the spring relaxed with the Casimir force decreasing. A simple diode rectifier in the external circuit will rectify this alternating current (AC) and produce usable direct current (DC) electricity. Of course, efficiency calculations and measurements are needed to ensure the overunity, commercially viable production of energy.

An example of a similar but more sophisticated system with two parallel semiconducting plates separated by an variable gap that utilizes several concepts referred to above is Dr. Pinto's "**optically controlled vacuum energy transducer**." By optically pumping the cavity with a microlaser as the gap spacing is varied, "the total work done by the Casimir force along a closed path that includes appropriate transformations does not vanish..." This indicates that work or energy output is theoretically predicted for his ZPE converter based on the accepted laws of physics. Further discussion is in the "Casimir Engine" section of this chapter.

Forward to the Future

Dr. Robert Forward, who passed away in 2002, said,

"Before I wrote the paper everyone said that it was impossible to extract energy from the vacuum. After I wrote the paper, everyone had to acknowledge that you could extract energy from the vacuum, but began to quibble about the details. The spiral design won't work very efficiently... The amount of energy extracted is extremely small... You are really getting the energy from the surface energy of the aluminum, not the vacuum... Even if it worked perfectly, it would be no better per pound than a regular battery... Energy extraction from the vacuum is a conservative process, you have to put as much energy into making the leaves of aluminum as you will ever get out of the battery... etc... etc...Yes, it is very likely that the vacuum field is a conservative one, like gravity. But, no one has proved it yet. In fact, there is an experiment mentioned in my *Mass Modification* paper (an antiproton in a vacuum chamber) which can check on that.[144] The amount of energy you can get out of my aluminum foil battery is limited to the total surface energy of all the foils. For foils that one can think of making that are thick enough to reflect ultraviolet light, so the Casimir attraction effect works, say 20 nm (70 atoms) thick, then the maximum amount of energy you get out per pound of aluminum is considerably less than that of a battery. To get up to chemical

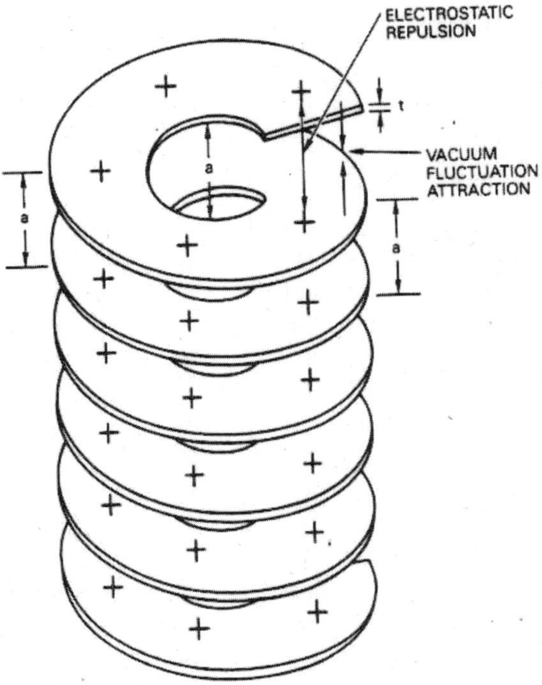

Figure 5.2 The very first journal article for converting zero point energy to electricity – Dr. Robert Forward called it a "spiral design for a vacuum-fluctuation battery" in 1984

energies, you will have to accrete individual atoms using the van der Waals force, which is the Casimir force for single atoms instead of conducting plates. My advice is to accept the fact that the vacuum field is probably conservative, and invent the vacuum equivalent of the hydroturbine generator in a dam."

To me this was very interesting advice, for someone like me just starting out in vacuum engineering in the 1990's. Forward is probably the first physicist to give considerable thought to the subject of this book, so he deserves some recognition for being a pioneer of ZPE applications. As it turns out, the spiral design is more of a battery than a generator. Also, the mounting evidence for the ZPF is weighing heavily in favor of being a non-conservative field. This means the experts are finding more ways to extract energy from ZPE without putting energy in first (see the next section for example).

Casimir Engine

The Casimir force presents a fascinating exhibition of the power of the ZPF offering about **one atmosphere of pressure** when plates are less than one micron apart. As is the case with magnetism today, it has not been immediately obvious, until recently, how a directed Casimir force might be cyclically controlled to do work. The optically-controlled vacuum energy transducer however, proposed by Fabrizio Pinto (a former Jet Propulsion Lab scientist), presents a powerful theoretical invention for rapidly changing the Casimir force by a quantum surface effect, excited by photons, to complete an engine cycle and thus transfer a few electrons. The convincing part of Pinto's invention is the quantum mechanics and thermodynamics that he brings to the analysis, offering a conclusive engine for free energy production. Pinto also worked with Robert Forward shortly after receiving his PhD and today, his company www.Interstellartechcorp.com is dedicated to "turning quantum vacuum engineering into a commercially viable activity."[145]

Pinto utilizes mechanical forces from the Casimir effect and a change of the surface dielectric properties, to intimately control the abundance of virtual particles. Basically, it is an **optically-controlled vacuum energy transducer**. Developed by the former Jet Propulsion Lab scientist, a moving cantilever or membrane is proposed to cyclically change the active volume of the chamber as it generates electricity with a thermodynamic engine cycle. The invention proposes to use the Casimir force to power the microcantilever beam produced with standard micromachining technology. The silicon structure may also include a microbridge or micromembrane instead, all of which have a natural oscillation frequency on the order of a free-carrier lifetime in the same material. The discussion will refer the (micro)cantilever design but it is understood that a microbridge or flexible membrane could also be substituted.

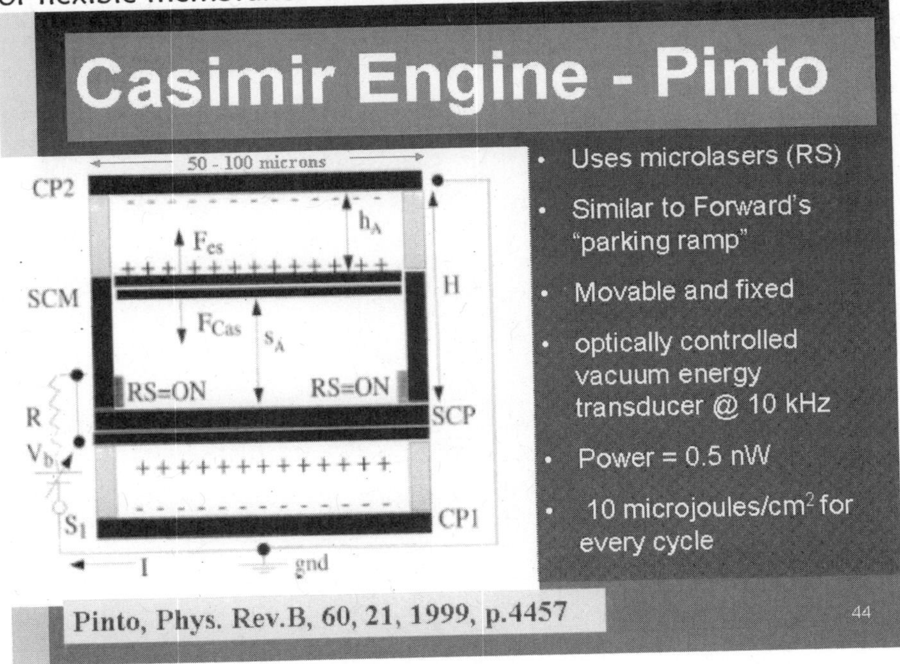

Fig. 5.3
Casimir Engine with a movable (upper) and fixed (lower) membrane design, 50-100 microns in size

The invention is based on the cyclic manipulation of the dimensions of Casimir cavity created between the cantilever and the substrate as seen in Fig 5.3. The semiconducting membrane

(SCM) is the **cantilever** that could be on the order of 50-100 microns in size with a few micron thickness in order to obtain a resonant frequency in the range of 10 kHz, for example.

Two monochromatic lasers (RS) are turned on thereby increasing the Casimir force by optically changing the dielectric properties of the cantilever. There is a frequency dependence of a dielectric constant which can vary with frequency by a few orders of magnitude but inversely proportional to the frequency. This means that as the frequency goes up, the dielectric constant may go down.

Pinto's pro-active approach is to excite a particular

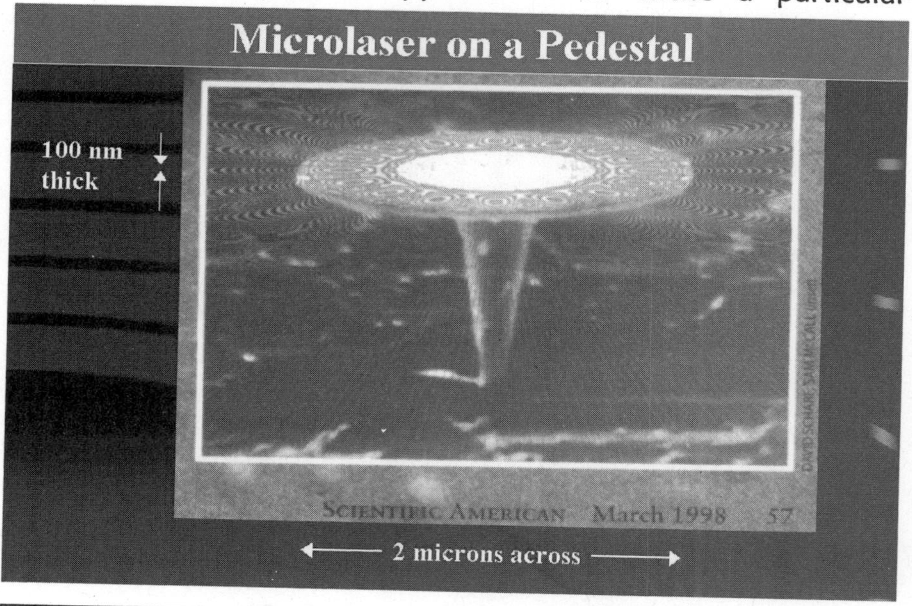

Fig. 5.4 Microdisk for use in Pinto's engine

frequency mode in the cavity. In doing so, an applied electrostatic charge (Vb) increases as the cantilever is pulled toward the adjacent substrate (SCP) by the Casimir force. Bending the charged cantilever on a nanoscale, the Casimir attractive force is theoretically balanced with opposing electrostatic forces, in the same way as Forward's "parking ramp" seen in Figure 5.2. As the potential difference to the cantilever assembly is applied with reference to a conducting surface (CP2) nearby, the distance to this surface is also kept much larger than the distance between the cantilever and the substrate (SCP). Upon microlaser illumination (Fig. 5.4), which changes the dielectric properties of the surface and increases the Casimir force, there is also predicted an increase in electrostatic energy due to an increase in capacitance and voltage potential. Therefore a finite electrical current can be extracted and the circuit battery is charged by an energy amount equal to the net work done by the Casimir force. Pinto estimates the Casimir force field energy transfer to be approximately 100 to 1000 erg/cm^2. Converting this to similar units used previously, this Casimir engine should produce in the range of 60 to 600 TeV/cm^2 (teraelectron volts per square centimeter) which is also approximately 0.01 to 0.1 mJ/cm^2 (millijoules per square centimeter or about **3.6 kWh/m^2**). This is a <u>significant energy density for a generator,</u> which can be upscaled as well.

Analysis of the Casimir Engine cycle demonstrates its departure from hydroelectric, gaseous, or gravitational systems. For example, the Casimir pressure always acts **opposite to the gas pressure of classical thermodynamics** and the energy transfer which causes dielectric surface changes "does not flow to the virtual photon gas."[146] Altering physical parameters of the device therefore, can change the total work done by the Casimir force, in contrast to gravitational or hydroelectric systems. Unique to the quantum world, the type of surface and its variation with optical irradiation is a key to the transducer operation. Normally, in classical physics, changing the reflectivity of a surface will affect the radiation pressure on the surface but <u>not the energy density of real photons.</u> However, in the Casimir force case, Pinto explains, "...the normalized energy density of the radiation field of virtual photons is drastically affected by the dielectric properties of all media involved via the source-free Maxwell equations."[147]

Specifically, Pinto discovered that **the absolute value of the vacuum energy can change** "just by causing energy to flow from a location to another inside the volume V." This finding predicts a major breakthrough in utilization of a quantum principle to create a transducer of vacuum energy. Some concerns are usually raised, as mentioned previously, with whether the vacuum energy is conserved. In quantum systems, if the parameters (boundary conditions) are held constant, the Casimir force is strictly conservative in the classical sense, according to Pinto. "When they are changed, however, it is possible to identify closed paths along which the total work done by this force does not vanish."

To conclude the energy production analysis, it is noted by Pinto that the frequency of 10kHz (10,000 cycles per second) is used as a performance limit. Taking the lower estimate of 100 erg/cm^2 per cycle, power or "wattage" is calculated to be about **1 kW/m^2 which is about the same as photovoltaic energy production.** However, this invention will work 24 hours a day, 7 days a week and is not dependent on the sun. The single cantilever transducer is expected to produce about 0.5 nW and establish a millivolt across a kilohm load, which is still fairly robust for such a tiny machine (Fig. 5.3).

The basis of the dielectric formula starts with Pinto's analysis that the "Drude model of electrical conductivity" is dependent on the mean electron energy <E> (less than hf) and estimated to be in the range of submillimeter wavelengths. The Drude model, though classical in nature, is often used for comparison purposes in Casimir calculations. The detailed analysis by Pinto shows that carrier concentrations and resistivity contribute to the estimate of the total dielectric permittivity function value, *which is frequency dependent*. The frequency dependence is of increasing concern for investigations into the Casimir effects on dielectrics. A higher frequency laser (such as green, blue or UV) may have significantly different effects than an infrared laser, for example.

Analyzing the invention for engineering considerations, it is clear that some of the nanotechnology necessary for fabrication of the invention has only become available very recently. The one-atom microlaser, invented in 1994, could be a key component for this invention since about ten photons are emitted per atom. However, it has been found that new phenomena, (1) the **virtual-photon tunnel effect** and (2) **virtual-photon quantum noise**, both have an adverse effect on the preparation of a pure photon-number state inside a cavity,[148] which may impede the performance of the microlaser if placed inside a cavity. Pinto points out that such a low emission rate is necessary since the lasing must take place "as a succession of very small changes."

Another suggested improvement to the original invention could involve a femtosecond or attosecond pulse from a disk-shaped semiconductor microlaser, such as those developed by Bell Laboratories (Fig. 5.4). The microlaser could be used in close proximity to the cantilever assembly. Such microlaser structures, called "microdisk lasers" measuring two microns across and 100 nm thick, have been shown to produce coherent light radially. An optimum choice of laser frequency would be to tune it to the *impurity ionization energy* of the semiconductor cantilever (impurity = doping element). In this example, the size would be approximately correct for the micron-sized Casimir cavity.

Pinto chooses to neglect any temperature effects on the "dielectric permittivity" (a measure of charge retention). However, since then, the effect of finite temperature has been found to be intimately related to the cavity edge choices that can cause the Casimir energy to be positive or negative.[149] Therefore, the contribution of temperature variance and optimization of the operating temperature seems to have become a parameter that should not be ignored. Also supporting this view is the evidence that the dielectric permittivity has been found to also depend on the change (derivative) of the dielectric permittivity with respect to temperature.[150]

The micro-fabrication task for the Casimir engine includes mounting microlasers inside the Casimir cavity and ensuring that an extended or continuous 10 KHz repetition rate is possible with a moving cantilever, for its expected lifespan. It is worth mentioning that similar cantilevers, made from silicon of the same micron size, with only one support end now operate in many new automobiles as acceleration and crash sensors, without high failure rates. The energy production rate for the Casimir engine is predicted to be fairly robust (0.5 nW per cell or **1 kW/m²**), which could motivate a dedicated research and development project in the future. However, the Casimir engine project of Pinto's appears to be a million-dollar investment at best, which is to be expected for such a complex, revolutionary invention. Pinto's business plan probably addresses the full-scale production costs and projected payback.

Utilizing some of the latest cavity QED techniques, such as mirrors, resonant frequencies of the cavity vs. the gas molecules, quantum coherence, vibrating cavity photon emission, rapid change of refractive index, spatial squeezing, cantilever deflection enhancement by stress, and optimized Casimir cavity geometry design, the Pinto invention may be improved substantially. The process of laser irradiation of the cavity for example, could easily be replaced with one of the above-mentioned quantum techniques for achieving the same variable dielectric and Casimir force effect, with less hardware involved. At the present stage of theoretical development, the Pinto device receives a moderate rating of feasibility. It's overall energy quality rating, in my opinion, is very high.

Rectifying Thermal and Non-Thermal (ZPE) Electric Noise

Diode rectifiers have been used since the early vacuum tube days to cause electricity in a circuit to go only one way. When solid-state semiconductors were invented, each tiny p-n junction replaced an entire vacuum diode tube. Recently, it has become more and more apparent that some diodes may be ideal for rectifying random noise from any source. For example, the U.S. Patent #3,890,161 by Charles M. Brown utilizes an array of **nanometer-sized metal-metal diodes**, capable of rectifying frequencies up to a terahertz (10^{12} Hz). Brown notes that thermal agitation electrical noise (Johnson noise) behaves like an external signal and can be sorted or *preferentially conducted in one direction by a diode*.

Johnson noise in the diode is also **generated at the junction itself** and therefore, requires no minimum signal to initiate the conduction in one direction. Brown's diodes also require no external power to operate. Brown also indicates that heat is absorbed in the system, so that a cooling effect is noticed, because heat (thermal noise) energy energizes the carriers in the first place and some of it is converted into DC electricity. In contrast, the well-known Peltier effect is the closest electrothermal phenomenon similar to this but normally requires a significant current flow into a junction of dissimilar metals in order to create a cooling effect (or heating). Brown suggests that a million nickel-copper diodes formed in micropore membranes, with sufficient numbers in series and parallel, can generate 10 microwatts. The large-scale yield is estimated to be **several watts per square meter**.

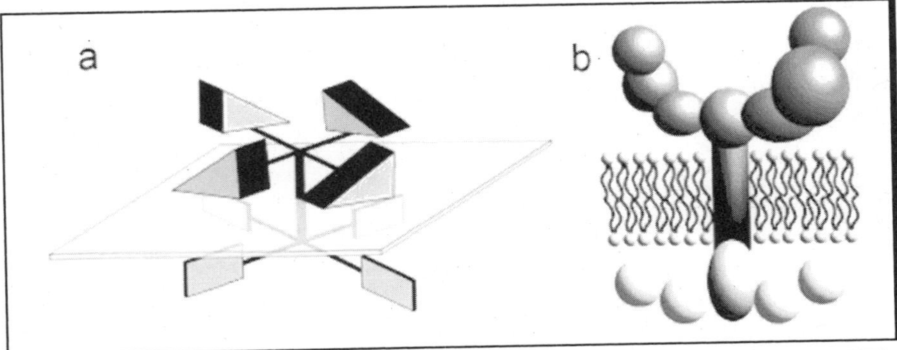

Figure 5.5 Molecular refrigerators that will transfer heat across an insulating barrier

Dr. Eric Davis, from the Institute for Advanced Studies in Austin, recently presented an update on the research they are conducting to prove that electron levels will change when put into a Casimir cavity.[151] It would be expected, from what the reader now understands as well, that inside a Casimir cavity, less frequencies (energy) are available. Consequently, if the electron levels in an atom are powered or bolstered by the virtual particle activity of the ZPF, they would tend to lower, or go below the ground state level for that atom, and emitting photonic radiation as it entered the cavity.

Instead of dismissing them as the US DOE does, let's look at molecules as a vacuum engineering nanotechnologist and see what some of them are doing today. Molecular diode rectifiers, or just microscopic zero-bias diodes, are the most fascinating to me. It may be true however, that nanodiodes or even

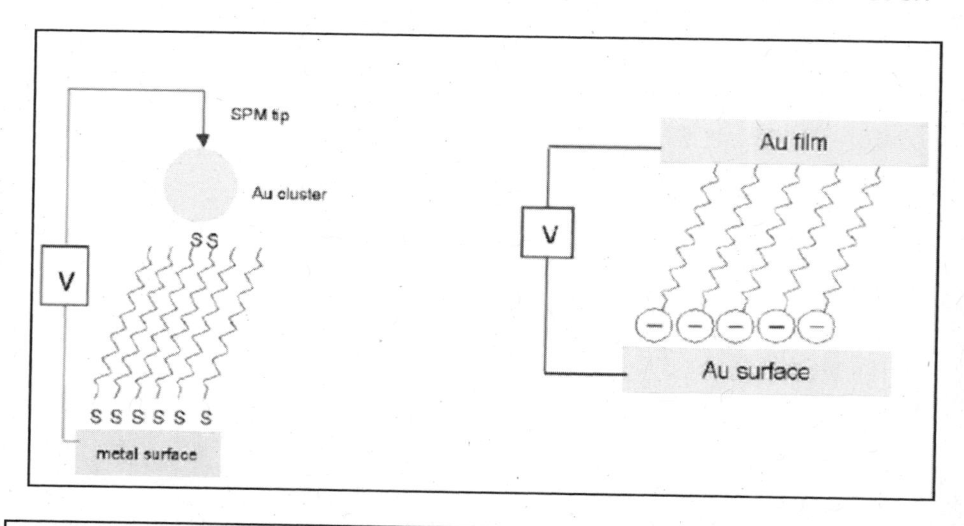

Figure 5.6 A scheme for converting ZPE non-thermal fluctuations into electricity, with Yu's self-assembled diodes.

semiconductor diodes can convert ZPE just as well. When the parameters that we are looking for are listed for the reader, many engineers will be able to go much further with this discovery of rectifying thermal and "**nonthermal noise**" which of course is the quantum vacuum ZPE noise that we want to use.

It is important to mention that physicists instead will be inclined to use liquid helium to cool any proposed non-thermal rectifier to near absolute temperatures, to prove that it will generate electricity only from ZPE. However, a much easier and more expedient path is to use the same bank of diodes (in parallel) to produce electricity from the fluctuations of thermal AND non-thermal energy to solve the planet's energy crisis right now.

Nanorefrigerators and Molecular Diodes

Environmentally, in this age of global warming, what better device to have in every home and business than **a solid-state, self-sustaining refrigeration unit that generates electricity?** That is exactly what has been re-discovered by Professor Van den Broeck from Hasselt University in Belgium.[152] In Fig. 5.5 we see the diagrams from his *Physical Review Letters* article that shows how such molecules will work. He describes them as a paddlewheel type of machine (Fig. 5.5a), which the real molecular model simulates (Fig. 5.5b). The exciting prediction is that a small thermal gradient will cause them to work or a mechanical force (0.1 piconewtons) will tend to create refrigeration.

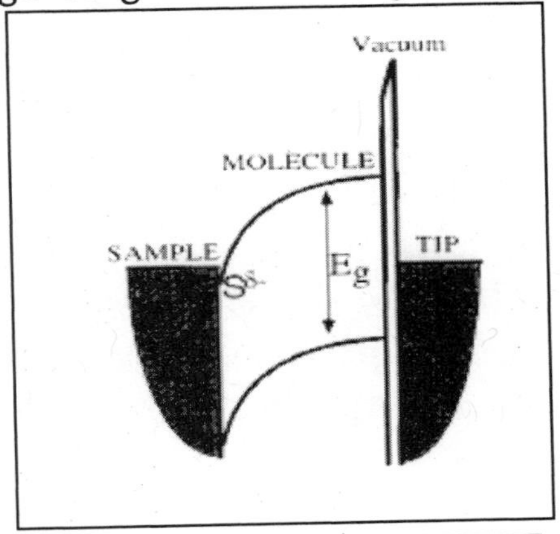

Fig. 5.7
Bandgap diagram for self-assembled diodes

When I read about this theoretical device being published, I was very gratified. It has been reported anecdotally that some energy generators, such as the patented Magnetic Energy Converter (MEC, see Chapter 2), does act as a refrigerator while in operation. It is logical that when converting ZPE, thermal energy will also be converted to electricity as well, causing a loss of heat in the system (causing cooling). Therefore, to have an independent verification of such a concept only strengthens the conclusion that this book is based upon: *solid-state ZPE converters are in our near future.*

It is suggested that the "most likely to succeed" ZPE converter is the rectifying diode adapted for the frequencies of interest, operating at zero bias in a backward mode. In electronics, *diodes* or "rectifiers" were direct outgrowths from the vacuum tube industry of our parent's generation. Some of us remember televisions made from vacuum tubes. The vacuum diode, with two electrodes, was the first tube to be manufactured, which conducted electrical current in only one direction. Then some smart engineer decided to put a variable-voltage screen in the midst of the two electrodes and call it a "triode" which had three electrodes and was the predecessor of the modern transistor.

The important lesson here, for the free energy fans in the audience, is the overlooked fact that semiconductor diodes have a **built-in power supply**. It is called the "junction potential" or "built-in potential" which creates a voltage between the positive (p) material and the negative (n) material. For our purposes, there are a certain class of diodes which do not create a barrier that requires a "forward bias voltage" to operate. (Normally silicon diodes need 0.7 volts before they can conduct and germanium needs 0.4 volts.) They are called **"zero-bias" diodes**. These are the key to transducing the thermal and non-thermal fluctuations right in their midst. In electronics, the fluctuations of either type are called "noise" because of their annoying characteristic "hiss" sound in amplifier circuits. However, it is music to my ears.

In regards to rectifying thermal electrical noise, the U.S. patent which started me thinking in this vein is #3,890,161 by Charles M. Brown that utilizes an array of nanometer-sized metal-metal diodes, mentioned previously. Brown, who I met in Portland while speaking at a New Energy Movement conference, notes that the diodes may also require no minimum signal to initiate the conduction in one direction, operating in a passive mode, with no external power needed. Along this line of thinking but on a smaller scale (for higher energy density) might be a plan for utilizing Dr. Yu's self-assembled diodes diagrammed in Fig. 5.6. In comparison to the standard semiconductor diodes we have discussed previously, these **self-assembled molecular diodes** have many features, such as being extremely tiny (nanometer-sized) and lining up like soldiers, which make them easy to work with.[153] Fig. 5.7 shows the band gap diagram (voltage bias barrier the electron must overcome) for these diodes, which has unique features. The vacuum barrier is very prominent in the figure, showing what amounts to a "tunneling" requirement for the electrons to go through the barrier, rather than over it. (Tunneling is perfectly allowable in quantum mechanics as long as the barrier is short.) This long chain molecule's zero bias resistance is around 100 gigaohms, which is very high. However, we recall from the previous chapter that Koch did his original work with ZPE tunneling in Josephson junction devices and Capasso notes the ZPE tunneling capability necessary for his patented solid state device.

Custom Made Zero Bias Diodes

Other diodes which exhibit the ability to rectify EMF energy include the class of "backward diodes" which operate with **zero bias** (no external power supply input).[154] These have been used in microwave detection for decades and have never been tested for nonthermal fluctuation conversion. There is every reason to presume they include such ZPE radiation conversion in their everyday operation but it is unnoticed with other EMF energy being so much larger in amplitude. US Patent 6,635,907 from HRL Laboratories describes a diode with a very desirable, "highly nonlinear portion of the I-V curve near zero bias." These diodes produce a significant current of electrons when microwaves in the gigahertz range are present. Another example is US Patent 5,930,133 from Toshiba entitled, "Rectifying device for achieving a high power efficiency." They use a tunnel diode in the backward mode so that "the turn-on voltage is zero." Could there be a better device for small voltage ZPE fluctuations that don't like to jump big barriers?

A completely passive, unamplified zero bias diode converter/detector for millimeter (GHz) waves was developed by HRL Labs in 2006 under a DARPA contract, utilizing an Sb-based "backward tunnel diode" (BTD). It is reported to be a "true zero-bias diode" that does not have significant 1/f noise when it is unamplified. It was developed for a "field radiometer" to "collect thermally radiated power" (in other words, 'night vision'). The diode array mounting allows a feed from horn antenna, which functions as a passive concentrating amplifier. The important clue is the "noise equivalent power" (NEP)of **1.1 pW per root hertz** (picowatts are a trillionth of a watt, so this is only one picowatt per every square root of the frequency in Hertz, which is very, very small) and the "noise equivalent temperature difference" of 10K, which indicate a sensitivity to Johnson noise, the source of which is ZPE.[155] Perhaps HRL Labs will consider adapting the invention for passive zero-point energy generation.

Another invention developed in 2005 by the University of California Santa Barbara is the "semimetal-semiconductor rectifier" for similar applications, to rival the metal-semiconductor (Schottky) diodes that are more commonly known for microwave detection. The zero bias diodes can operate at room temperature and have a NEP of about 0.1 pW but a high "RF-to-DC current responsivity" of about 8 A/W (amperes per watt). Most importantly, the inventors claim that the new diodes are about 20 dB more sensitive than the best available zero-bias diodes from Hewlett-Packard.[156]

There also have been other inventions such as "single electron transistors" that also have "the highest signal to noise ratio" **near zero bias**. Furthermore, "ultrasensitive" devices that convert radio frequencies have been invented that operate at outer space temperatures (3 degrees above zero point:

3°K).[157] These devices are tiny nanotech devices so it is possible that lots of them could be assembled in parallel (such as an array) to produce ZPE electricity with significant power density.

Dr. Peter Hagelstein from Eneco, Inc. was thinking along the same lines when in 2002 he patented his "**Thermal Diode for Energy Conversion**" (US Patent 6,396,191) which uses a thermopile bank of thermionic diodes. These are slightly different, more like thermocouples, than the diodes that I am advocating. However, Hagelstein's diodes are so efficient that he predicts that, with only a 10°C temperature difference, a water pool of six meters on a side could supply the electricity for a house. He also suggests their use as "efficiency boosters" for augmenting the performance of electric or hybrid cars.

Other devices which also will provide the fueless electrical energy cars, planes and homes need simply use zinc oxide or titanium oxide films that can convert ambient heat into electricity, as used in photovoltaic panels. A few reports indicate that these work reliably for years. Such solid-state diode converters will also grab the nonthermal ZPE in the process and therefore can work in outer space, with no sun.

Summary

"At least one mechanism for the continuous extraction of limitless amounts of 'free energy' from the vacuum" includes trapping antiprotons and monitoring their energy-releasing annihilation, which Dr. Forward suggests may be random.[158] Another mode of conversion of ZPE energy is with high voltage discharges, which also have anomalous, overunity energy outputs. A recent physics analysis of lightning also confirmed the apparent overunity production of energy present in a thundercloud bolt discharge. Sufficiently high electric field gradients cause the vacuum to breakdown (or decay) into particle pair production of energy, which is seen as the "polarization" of the vacuum near the boundaries of charged particles and also the voltage discharge phenomenon.[159] Dr. Peter Graneau has demonstrated a high voltage discharge energy production of 20 Joules and attributes it to "chemical bond energy."[160] However, it is also known that orbital and bond energy can be traced to ZPE.

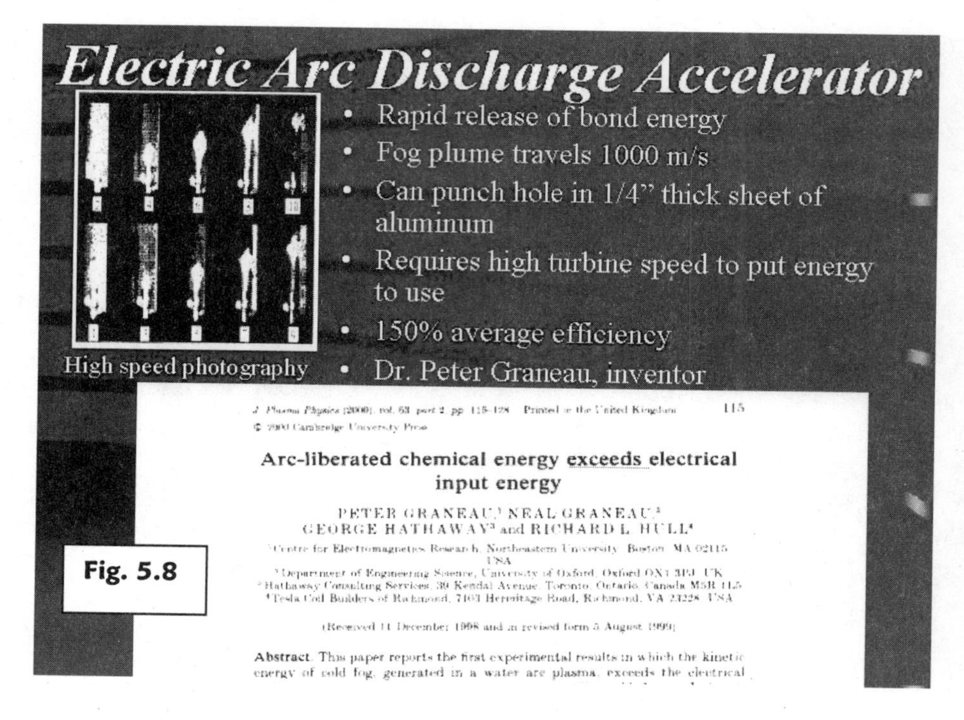

Fig. 5.8

It can be easily concluded that there is a clear pathway for ZPE to be the sustaining energy source for all energy states of the atom, including the angular momentum **J** of the electron. With that physics principle established, the macroscopic magnetic field of a permanent magnet, which is totally attributed to the **J** of the electron, can be said to be sustained by the vacuum ZPE field. Therefore, as entirely **permanent magnet motors**, generators, and actuators become commercialized, it will no longer be a mystery as to where the energy is coming from. Furthermore, these magnet-powered devices cannot be mistaken for *perpetuum mobiles*. (excerpt from "Are Magnetic Fields Connected to ZPE?" in Appendix).

The photovoltaic compounds mentioned in this chapter are already in use but haven't been engineered for non-solar, thermal conversion. The zero bias diode can operate with any miniscule amount of electrical energy, which is exactly what Johnson noise is. It is just a matter of will power and initiative to begin producing such products on a large scale, once a target power density is selected. With thousands of sub-millimeter-sized diodes that are assembled for the consumer market of flat screen monitors and televisions, it is reassuring to know that the same assembly process will work for ZPE diode generators. Furthermore, as the Brown patent suggests, they can be stacked for a maximum energy density of perhaps hundreds of kilowatts per cubic meter. Then the promise of compact, stand-alone distributed generators in every home and business will be come a reality in our lifetime.[161]

ADVANCED TECHNOLOGY

Aviation Week & Space Technology, March 1, 2004

To the Stars

Zero point energy emerges from realm of science fiction, may be key to deep-space travel

WILLIAM B. SCOTT/AUSTIN, TEX.

At least two large aerospa
panies and one U.S. I
Dept. agency are betti
"zero point energy" coul
next breakthrough in ae
vehicle propulsion, and are l
those bets with seed money for
search.

If their efforts pay off, ZPE
powerplants might enable Mach
ers, quiet 1,200-seat hypersoni
ers that fly at 100-mi. altitudes
12,000 mi. in about 70 min., and
trips to the Moon.

ONE OF THOSE companies, B
tems, launched "Project Green
1986 "to provide a focus for resea
novel propulsion systems and th
to power them," said R.A.
Evans, the project leader, in a
technical paper last year. Although funding levels have
been modest, Greenglow is
exploring ZPE as one element

PE-relat-

energy is
d is diffi-
ıinded to
by metic-

Spacecraft capable of interstellar travel will approach the speed of light, and may have to extract energy from the vacuum of space. However, researchers could be years or decades from achieving the breakthroughs necessary to build such a propulsion system.

cowatts or
That st
searchers,
some crit
tion. Still,
ernment

Figure 6.1 Inset from *Aviation Week* magazine advocating ZPE for space travel

Chapter 6

Propulsion and Space Travel from ZPE

Overview

As a kid, I was enthralled with science fiction. Buck Rogers, Flash Gordon and Capt. Kirk were my first heroes of space travel. As I grew up and studied graduate physics, especially special and general relativity, I found that there has been very little scientific progress in the direction we sci-fi fans want to go. A decade ago, Alcubierre published a paper proving that the concept of a "warp drive" was actually physically sound (see review article in Appendix).[162] However, he showed

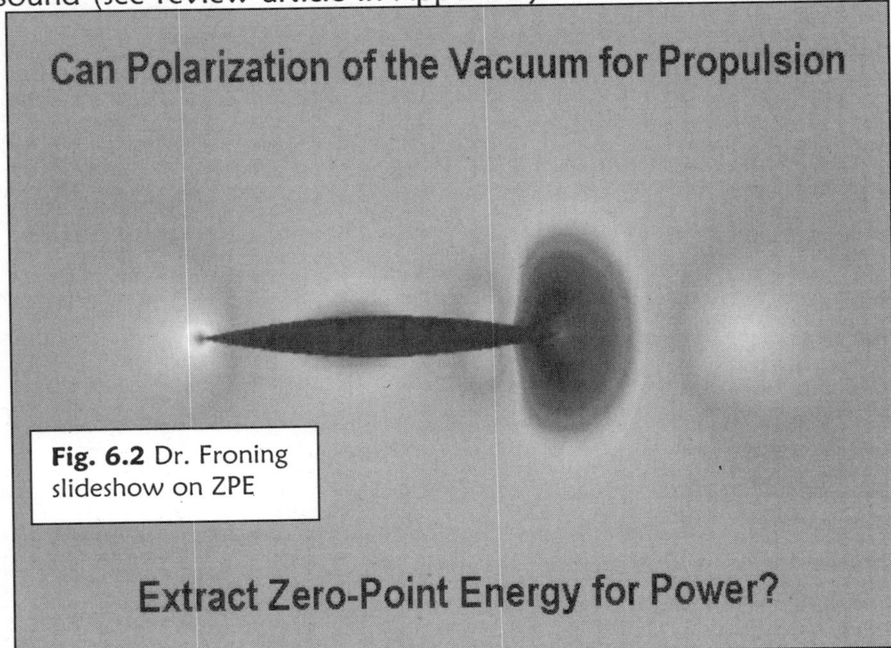

Can Polarization of the Vacuum for Propulsion
Extract Zero-Point Energy for Power?

Fig. 6.2 Dr. Froning slideshow on ZPE

that an enormous amount of energy was required to produce the warping of space to compress space-time in front of the ship and expand space-time behind it. Everyone chuckled with an "I told you so" and quickly forgot about the intellectual exercise. However, many of us believe that space travel is our future. In fact, I feel it is our "manifest destiny" just like when the early settlers started believing in the coast-to-coast view of the United States. Furthermore, the ZPF is the only reservoir that holds enough energy for such a process to work. We just need to learn how to take advantage of it to achieve our destiny.

Arthur C. Clarke, with whom I have corresponded, once said,

> ***"The earth is the cradle of civilization but mankind cannot live in the cradle forever."***

Clarke is known for many prophetic

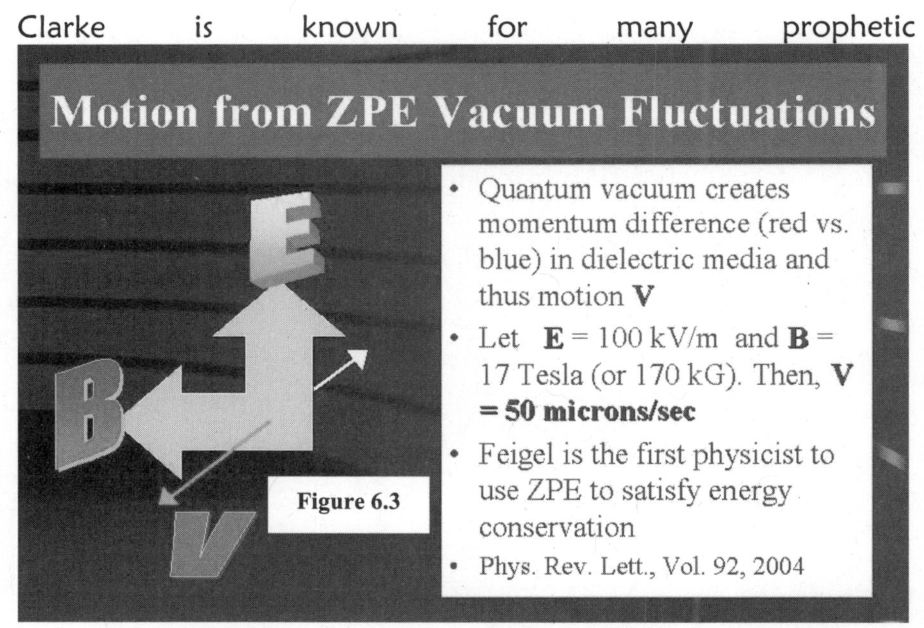

Motion from ZPE Vacuum Fluctuations
E
B
V
• Quantum vacuum creates momentum difference (red vs. blue) in dielectric media and thus motion **V**
• Let **E** = 100 kV/m and **B** = 17 Tesla (or 170 kG). Then, **V = 50 microns/sec**
• Feigel is the first physicist to use ZPE to satisfy energy conservation
• Phys. Rev. Lett., Vol. 92, 2004

Figure 6.3

technical achievements and in his book, *3001*, he predicts that the Haisch, Rueda and Puthoff (HRP) "inertialess drive" will most likely be put to use like a controllable gravity field, thanks to the landmark paper by Haisch et al.[163] Clarke writes in this book, "...if HR&P's theory can be proven, it opens up the prospect—however remote—of antigravity 'space drives,' and the even more fantastic possibility of **controlling inertia.**"[164] This prediction is compatible with the new theoretical nature of inertia. It will be just a matter of time for the inertialess EM shield to be designed.

ZPE Experimental Propulsion

ZPE has evolved from years of speculation and theory to amazing effects in the lab, which are quite numerous. Fabrizio Pinto, whose company Interstellar Technologies and published journal papers have demonstrated a commitment to the quantum vacuum revolution, has stated, "Fuel-free propulsion does not violate any conservation laws."[165]

In 2004, **Alexander Feigel** proposed that the *net momentum of the virtual photons* can depend upon the direction in which they are traveling, if they are in the presence of electric or magnetic fields. His theory and experiment offers a possible explanation for the accelerated expansion of distant galaxies, as well as the motion of fluids.[166] Furthermore, Feigel's discovery opens the way to harness virtual particles to do useful work by creating propulsion (see Fig. 6.3).

Its fundamental influence on heat, inertia, and gravity can be found in Puthoff's recent paper entitled, **"Engineering the Zero-Point Field and Polarizable Vacuum for Interstellar Flight."**[167] In it he states,

> One version of this concept involves the projected possibility that empty space itself (the quantum vacuum, or space-time metric) might be manipulated so as to provide energy/thrust for future space vehicles. Although far-reaching, such a proposal is solidly grounded in modern theory that describes the vacuum as a polarizable medium that sustains energetic quantum fluctuations.[168]

A similar article proposes that "monopolar particles could also be accelerated by the ZPF, but in a much more effective manner than polarizable particles."[169] Furthermore, "...the mechanism should eventually provide a means to transfer energy...from the vacuum electromagnetic ZPF into a suitable experimental apparatus."[170] With such endorsements for the propulsion use of ZPE, the value of this present book seems to be validated and may be predicted to yield scientific results which will be quite fruitful.

Jet Propelled to the Future

As the keynote speaker for our institute's Second International Conference On Future Energy (COFE), Dr. Fabrizio Pinto, a former Jet Propulsion Lab physicist, presented his research into applications of ZPE. Pinto is famous for his Casimir Engine, which converts ZPE into electricity.[171] One of the findings he presented is an explanation of two ships in relative proximity to collide, even if they are in still water. Apparently this phenomenon is well known among sailors and Dr. Pinto has experienced it as well. The dissimilar wavelengths on either sides of the boats are just like the Casimir force, as described in the book *Van der Waals Forces*.[172] Another discovery he notes,

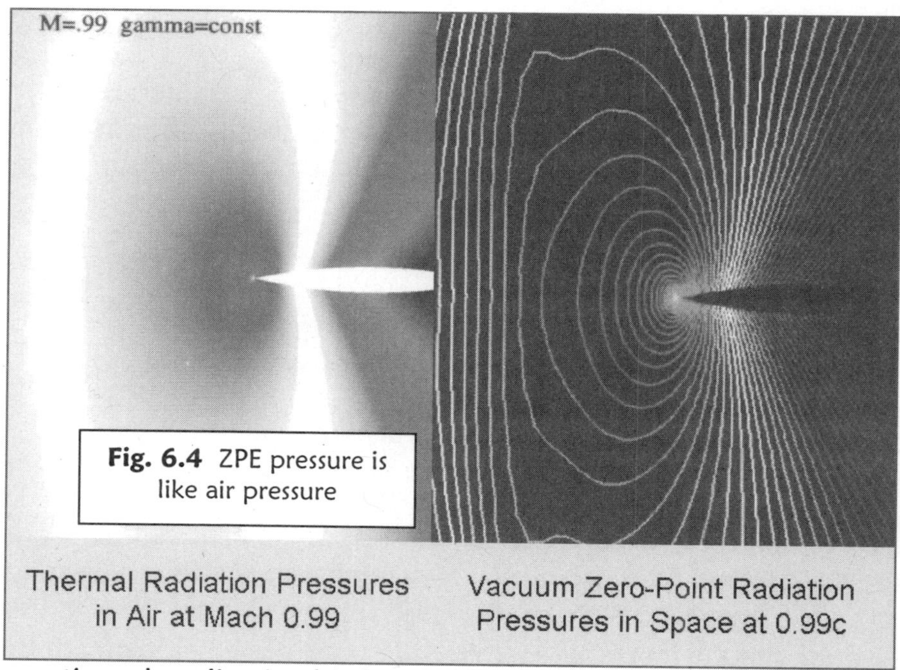

Fig. 6.4 ZPE pressure is like air pressure

mentioned earlier in this book, is how geckos are able to cling to any surface because their paws' tiny hairs are small enough to create a Casimir attraction when touching a flat surface. It turns out that virtually all insects also use the Casimir force to cling to any surface. The *Van der Waals Forces* text calculates that even an insect 2 cm across could cling to the ceiling if it had 4 cm^2 of contact surface in total and a heavy density close to that of

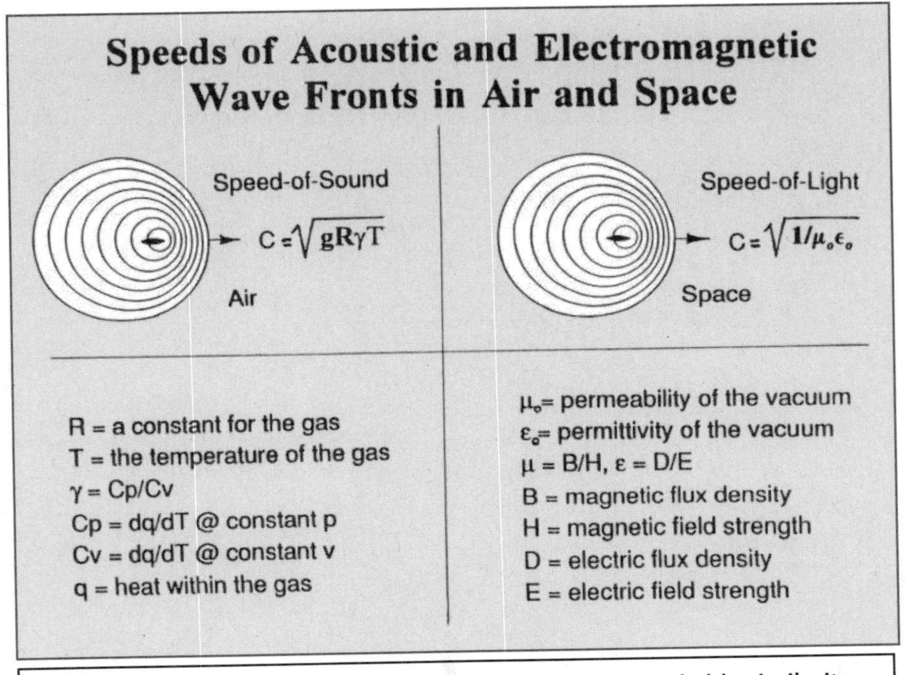

Speeds of Acoustic and Electromagnetic Wave Fronts in Air and Space
Speed-of-Sound
$C = \sqrt{gR\gamma T}$
Air
Speed-of-Light
$C = \sqrt{1/\mu_o \epsilon_o}$
Space
R = a constant for the gas
T = the temperature of the gas
γ = Cp/Cv
Cp = dq/dT @ constant p
Cv = dq/dT @ constant v
q = heat within the gas
μ_o= permeability of the vacuum
ϵ_o= permittivity of the vacuum
μ = B/H, ε = D/E
B = magnetic flux density
H = magnetic field strength
D = electric flux density
E = electric field strength

Fig. 6.5 Comparison of light and sound shows remarkable similarity

water. However, insects' density is much lower than water and therefore, their contact area can be much lower in area. Pinto also showed, during his COFE presentation, a translated article of Enrico Fermi, published in 2006, that demonstrates a classical interaction of charges with gravity with "drooping electric field lines." Pinto's opinion of this article was that "**Gravity alters the laws of electrostatics and electromagnetics**." He also has interpreted this discovery as a lifting means (antigravity) for propulsion in *Journal of Physics D.*[173] Dr. Pinto also has discovered a Casimir-based particle propulsion method.[174] A DVD of his educational presentation, as well as a "Collected Works of Fabrizio Pinto," is available from IRI.[175]

ZPE for Space Travel

While at the Utah Chapter of the National Space Society in 2006, I had the pleasure of meeting Dr. David Froning, who has published numerous articles on ZPE and the Casimir force. In particular, we talked about his ongoing interest in space travel utilizing "faster than light" (FTL) ZPE discoveries of his. In Fig. 6.1, there is a sample of an article from *Aviation Week and Space Technology* that proposes such a possibility. Furthermore, after meeting and talking with Froning, I am much more convinced of its rationality. Though the US Patent Office presently has a ban on such inventions (FTL is in the Sensitive Applications Warning System), Dr. Froning shows there is a lack of substantial difference between the surmountable speed of

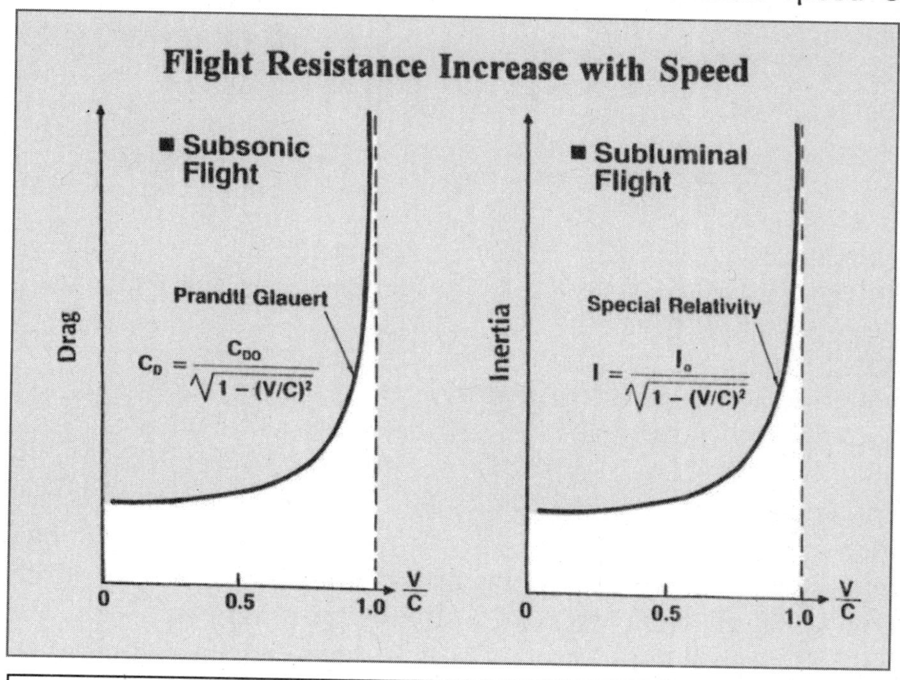

Fig. 6.6 Correlation of light and sound speed formulas

sound and the "insurmountable" speed of light. Note how (Fig. 6.5) only the electromagnetic characteristics of the ZPF are the primary ingredients of its permeability (μ) and permittivity (ε) in the speed of light formula. (Permeability can be defined as how "permeable" the substance is to any magnetic flux, where iron is more permeable than air.) These correspond to or are analogous to the gas constant, pressure, volume and temperature in the speed of sound equation.

Figs. 6.5-6.6 and 6.7-6.8 offer an amazing insight between the equations for the speed of sound and the speed of light. After studying both of these figures, we should be considering the possibility that the experts might be making a mistake in letting Einstein control our future. To me, it is only a matter of time before Froning starts demonstrating small-scale experiments that exhibit a breakthrough of the light barrier, because he knows how to reduce the drag of the vacuum!

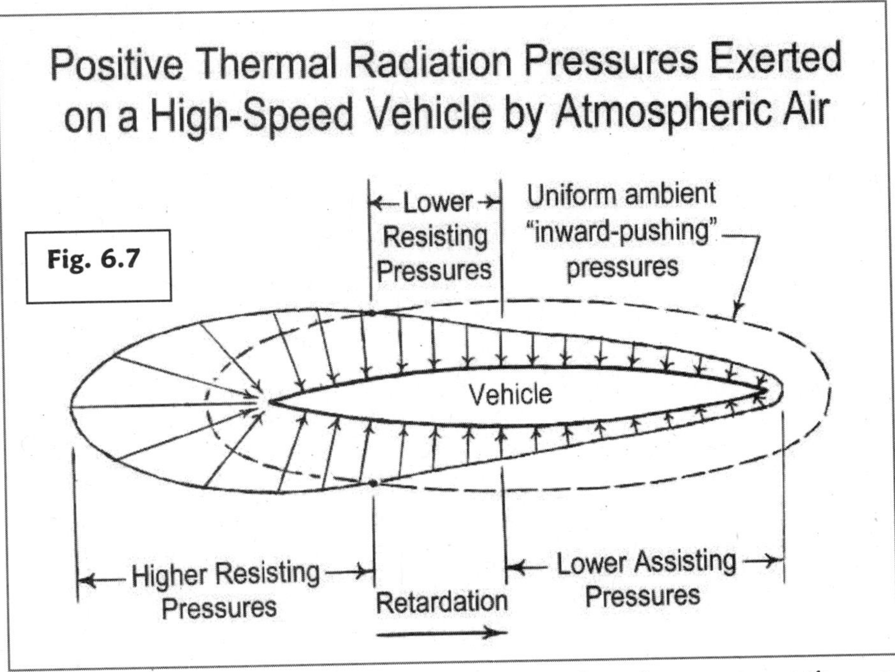

The question of drag or inertia for the propagation of light and/or any material body in space has always been a mysterious question in physics. The only satisfactory answer that the Standard Model provides for this problem is "Mach's Principle" which Einstein relied upon a century ago. It states that the distance stars provide a reference frame so that when turning or accelerating, there is some semblance of an explanation for the resistance that space itself presents. Instead, Puthoff, Rueda and Haisch's work dramatically presents a much more satisfactory answer, based on the ZPF.[176]

Figs. 6.7 and 6.8 offer more details about the comparison of the dynamic air pressure on the same slim, cross sectional design of a vehicle. Note the important difference that his simulation predicts in the assisting pressures and the resisting pressures produced simply by radiation. In both figures, the vehicle is moving to the left. The conclusion is that *positive* thermal radiation pressure is exerted on the leading edge of the craft that resists its forward motion. However, in the ZPE situation out in space, the vacuum behaves in the opposite

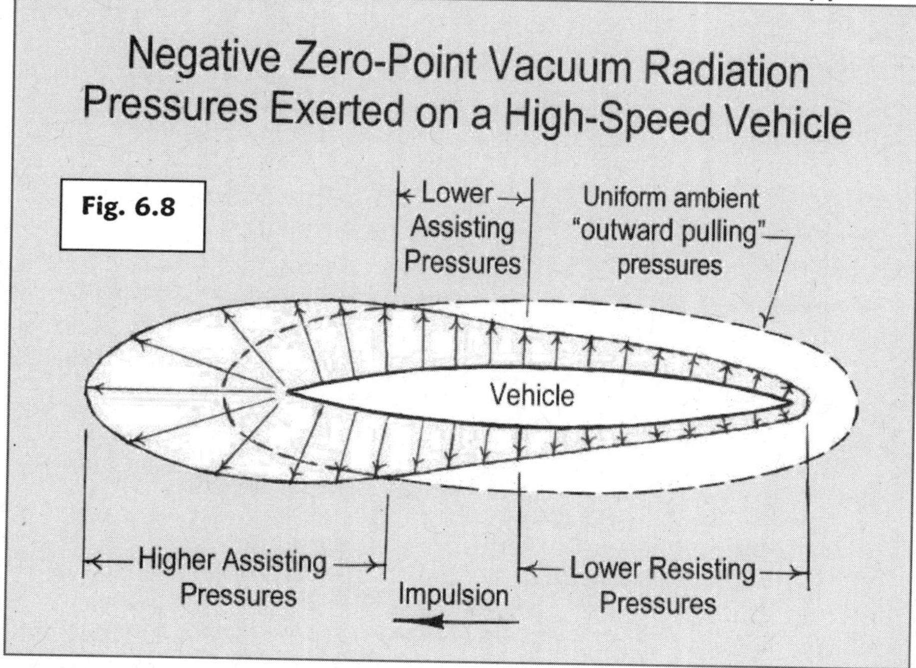

fashion. The *negative pressure on the leading edge actually assists the forward motion*, especially if the electromagnetic field surrounding the ship is increased, such as with a toroid.

Froning proposes to do this equally by creating a "vector potential" (A) which is a fancy name for a high current, donut-shaped coil surrounding the entire vehicle. Increasing the vector potential will tend to perturb the permeability and permittivity in the local space surrounding the vehicle accompanied by a distortion of the ZPF. Their approach is particularly to use alternating current toroids with resonant frequencies. Without going into the physics too much, the peer-reviewed physics journals like his proposal.

In Fig. 6.9, we see a drawing of the craft and the surrounding disturbance (called by physicists a "perturbation") of the ZPF, which is assisting the transport of the vehicle with this enhanced toroidal vector potential field.[177] According to **Froning and Roach**, the physical representation of the craft can be changed by surrounding a saucer-shaped spaceship with a toroidal EM field that distorts and perturbs the vacuum

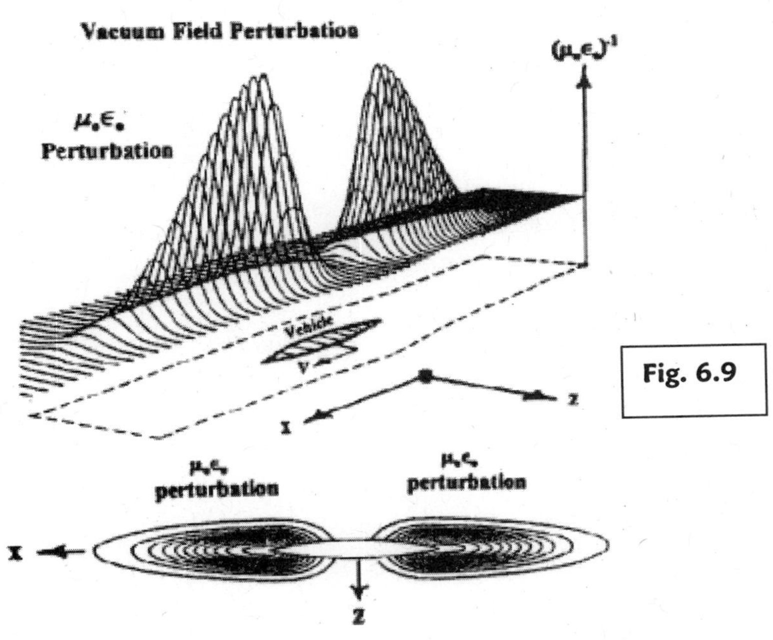

Fig. 6.9

sufficiently to affect its permeability and permittivity. The vacuum field perturbations are simulated by fluid field perturbations that resulted in the same percentage change in disturbance propagation speed within the region of perturbation. The computational effort was simplified by solving only the Euler equations of *fluid dynamics* for wave drag, used in hydrodynamics. The resulting μ_o and ε_o perturbation solutions are graphically shown in Fig. 6.9.

Casimir Cavity Opposes a Gravitational Field

One of the more obscure discoveries in recent years relates to another form of true *antigravity*. Dr. Pinto reports on this effect in his paper on quantum vacuum propulsion.[178] By accounting for the effect that spacetime curvature has on the modes of oscillation of the electromagnetic fields inside a Casimir cavity of two parallel plates at rest and parallel to the ground, an "antiparallel" net self-force appears that opposes gravity. The force is equal to the gravitational equivalent mass of the Casimir energy in the cavity. The equation for the force has a negative sign, indicating a small but significantly nonzero lifting force![179]

Summary

While there are many articles describing general approaches to utilizing the ZPF for propulsion, it is important to keep track of the physical breakthroughs that are cited for evidence this is really possible.[180] It has been proposed by Rueda and Haisch for example, with a contract from NASA, that *the ZPF can lose its* "**Einstein-Hopf drag**" *as the absolute temperature approaches zero, which will leave only the accelerating recoil force left.* (The temperature of outer space is about 3K, or 3 degrees above absolute zero, so this finding is very pertinent for space travel.) Furthermore, they propose that the ZPF can provide a directional acceleration to monopolar particles more effectively that to polarizable particles. They also suggest that "if valid, the mechanism should eventually provide a means to transfer energy, back and forth, but most importantly forth, from the vacuum electromagnetic ZPF into a suitable experimental apparatus."[181] With self-consistent theory and experiment, ZPE space propulsion is headed for the near future, perhaps by 2012.[182]

Below we find that more and more companies are realizing the possible bonanza that may be found in the near future with a little research in any one of the directions available for ZPE research and development. (Thanks to Gene for offering this information.)

-----Original Message-----
From: Genebazan@aol.com [mailto:Genebazan@aol.com]
Sent: Wednesday, January 03, 2007 8:38 PM
To: iri@erols.com
Subject: ZPE

Dear Tom and Jackie,

Best for the New Year!

My old consulting partner, Dave Allen, is VP for Technology Transfer at the Univ. of Colorado. He stopped by to visit for a day during the holidays.

His office gave a small grant of $25,000 to Garret Moddel, Electrical and Computer Engineering, Here is the citation for Moddel from their Annual Report (p. 8): "Production of Hydrogen via Water Splitting. Envisioned is a nanotechnology device that constrains individual atoms to the extent that zero point energy can be captured."

Dave went on to comment that Moddel got $800k from DARPA for ZPE.

Best,
Gene Bazan

Fig. 7.1 Buildings will be powered by ZPE generators as the technology scales up

Chapter 7

How ZPE Products Will Affect Us

Overview

As any new, disruptive technology is introduced into society, the scramble to adapt can be significant. Just ask any venture capitalist. However, if ZPE products start to emerge in handheld and computer backup-battery replacements, the change may not be noticed for a while. This often is a technique, called "stealth marketing" to delay the competition from starting to catch on to the new trend. As groups like the World Future Society (www.wfs.org), Kurzweil (www.kurzweilai.net), and Arlington Institute (www.arlingtoninstitute.org) predict, the future will offer much more technology that interfaces with the body and brain. What better way to spur the development of electronic technology than to have **self-powered devices**? We already have it in the form of solar-powered calculators, wristwatches that don't need battery changes, and the BetaBatt Company which is developing a 25-year tritium battery, to give you three examples.

The megatrend has begun: more and more products now come with "energy for a lifetime." The integration of ZPE products into society will kick it up another notch and be even more sudden, disruptive, and discontinuous. It has to be. When your neighbors catch wind of bigger and larger scale consumer items that have a mysterious power cell that needs no fuel, stock prices for those companies will soar as they rush to be the first to invest. A "one-time investment" in any and every product that does not need to be plugged in or filled up means the consumer will want to buy and hold onto that one over any other like it. Third world countries can be transformed into useful contributors to society with a single investment in or donation toward their own ZPE power cell for the community or household.

Future ZPE Developments Will Create Freedom

As small devices are introduced to the market, which is the most likely niche to fill, such as solar-powered calculators, consumers will notice that *a one-time investment creates a longer lasting, more stable, higher standard of living*. Have you every noticed how quickly our Western standard of living disappears as soon as an inconvenient weather disaster hits? Let a power line go down and millions can be forced back into primitive living with fire and flashlights in an instant. Our present modernization is quite unreliable, being dependent on a long umbilical cord of centralized power. Of course, the new ZPE future will come at a price. However, as solid-state integration of nanotechnology becomes more and more inexpensive, we will find a decline in energy-demanding costs. It is worth noting that in many new cars, a tiny chip now contains a *nanotech swinging weight to detect collisions*. These are accelerometers (acceleration sensors) that are very reliable. It just goes to show how even complex nano-devices like Pinto's Casimir Engine will someday become commonplace.

Fig. 7.2
Proposed Excalibur 1000-Watt ZPE generator

In Fig. 7.2, we see a model of a kilowatt generator that is proposed to extract ZPE for electricity supply. Having units like these for the home, car, and office will make everyone's lifestyle entirely different than today. It is a guaranteed upgrade in the standard of living for any third world country and a great standard of freedom for explorers, astronauts and military personnel.

Recently, Dr. Richard Obousy, a ZPE investigator, said,

> From the papers studied the author has grown increasingly convinced as to the relevance of the ZPE in modern physics. The subject is presently being tackled with appreciable enthusiasm and it appears that there is little disagreement that the vacuum could ultimately be harnessed as an energy source. Indeed, the ability of science to provide ever more complex and subtle methods of harnessing unseen energies has a formidable reputation. Who would have ever predicted atomic energy a century ago?[183]

We may therefore wonder about society's issues that are preventing ZPE acceptance and development.

Perspective on a ZPE Society

An investor and entrepreneur sent an email to me recently about his progress with an invention that is exhibiting the characteristics we associate with those that tap the ZPF. His comments are valuable to gain a perspective on the state of society after his product reaches the market:

> Posit if you will, a device that does the following: generates 100 kW, enough to power a car or two houses, the size of 2 cubic feet; scalable to Megawatts in the size of a garage; miniature versions to self-power tools, appliances, virtually anything. It can replace car, truck, locomotive and ship engines.
>
> It generates this energy without consuming anything because it is a zero point energy transducer, so the costs for the electricity or motive force produced are those of maintenance; we guess about 0.1 cent per kWh, or less, a factor of 30 to 70 times cheaper than current fossil / atomic generation.
>
> No noxious products are produced either. This means that desalinization plants would be practical. There would be instant energy production available for third world countries to use in their development. Like cell phones, no copper infrastructure would be needed as all generation is done locally. These devices will not be expensive either. In developed countries, each home

> and business could have one of our generators and obsolete the centralized model of power distribution.
>
> Sounds good so far? Of course we can see the obvious crash and burn of the existing energy and power distribution companies, the shift in power away from oil producing countries and huge layoffs of personnel in these industries, but what of the longer term consequences?[184]

The response from the Arlington Institute was also informative and encouraging, giving the impression that many people are poised and ready, including the military, for the next energy breakthrough:

> Very nice to hear from you, ... I appreciate all of this. I'm quite interested in what you are talking about, both in terms of the underlying technology and in the potential implications of it. To that end, I've set up a venture fund with a couple of sophisticated partners to fund breakthrough technologies that could "change the world". We are standing that up right now and would be most interested in hearing in more detail about what you've got going. Part of the services that The Arlington Institute would provide for the fund would be studying the potential impact and implications of a target investment opportunity, so there is certainly a vehicle in place to do some of the thinking that you are interested in. Furthermore, we are soon to start a study for the Department of Defense to build a strategy for the US to accelerate our extrication from a dependence on oil and gas and move to the next major energy source. In that project we will be surveying all of the existing alternative energy source (both conventional and unconventional) that we can find -- so we will have a pretty good sense of what the potential energy landscape looks like. So I'd like to hear more.

Importance of a ZPE Breakthrough

It is unduly apparent that research into this ubiquitous energy is overdue. The question has been asked, "Can new technology reduce our need for oil from the Middle East?" More and more sectors of our society are demanding breakthroughs in energy generation, because of the rapid depletion of oil reserves and the environmental impact from the combustion of fossil fuels. "In 1956, the geologist M. King Hubbert predicted that U.S. oil production would peak in the early 1970s. Almost everyone, inside and outside the oil industry, rejected Hubbert's analysis. The controversy raged until 1970, when the U.S. production of crude oil started to fall. Hubbert was right. Around 1995, several analysts began applying Hubbert's method to world oil production, and most of them estimate that the peak year for world oil will be between 2004 and 2008. These analyses were reported in some of the most widely circulated sources: *Nature*, *Science* and *Scientific American*. None of our political leaders seem to be paying attention. If the predictions are correct, there will be enormous effects on the world economy." Figure 7.3 is taken from the Deffeyes book showing the Hubbert method predicting world peak oil production and decline.[185]

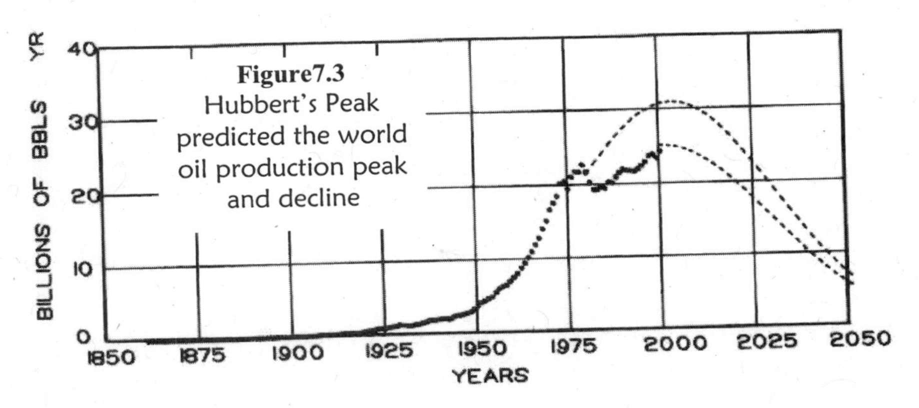

Figure7.3
Hubbert's Peak predicted the world oil production peak and decline

It is now widely accepted, especially in Europe where I participated in the World Renewable Energy Policy and Strategy Forum, Solar Energy Expo 2002 and the Innovative Energy Technology Conference, (all in Berlin, Germany), that the world oil production peak will probably only stretch to 2020, and that global warming is now occurring faster than expected. Furthermore, it will take decades to reverse the damage already set in motion, without even considering the future impact of "thermal forcing" which the future greenhouse gases will cause from generators and automobiles already irreversibly set in motion. The Kyoto Protocol, with its 7% decrease to 1990 levels of emissions, is a small step in the right direction but it does not address the magnitude of the problem, nor attempt to reverse it. "Stabilizing atmospheric CO_2 concentrations at safe levels will require a 60 to 80 per cent cut in carbon emissions from current levels, according to the best estimates of scientists." Therefore, renewable energy sources like solar and wind power have seen a dramatic increase in sales every single year for the past ten years as more and more people see the future shock looming on the horizon. Solar photovoltaic panels, however, still have to reach the wholesale level in their cost of electricity that wind turbines have already achieved.

Another emerging problem that seems to have been unanticipated by the environmental groups is that too much proliferation of one type of machinery, such as windmills, can be objectionable as well. Recently, the Alliance to Protect Nantucket Sound filed suit against the U.S. Army Corps of Engineers to stop construction of a 197-foot tower being built to collect wind data for the development of a wind farm five miles off the coast of Massachusetts. Apparently, the wealthy residents are concerned that the view of Nantucket Sound will be spoiled by the large machines in the bay. Therefore, it is likely that only a compact, distributed, free energy generator will be acceptable to the public in the future. Considering payback-on-investment, if it possessed a twenty-five year lifespan or more, while requiring minimum maintenance, then it will probably please most of the people, most of the time. The development of a ZPE generator theoretically would actually satisfy these criteria.

Compartmentalized and Classified Programs

Dr. Steven Greer of the Disclosure Project has stated,

> Classified above top-secret projects possess fully operational anti-gravity propulsion devices and new energy generation systems, that, if declassified and put to peaceful uses, would empower a new human civilization without want, poverty or environmental damage.[186]

However, the declassification of black project, compartmentalized exotic energy technologies is not readily forthcoming. Moreover, it is not even acknowledged. (The first step toward declassification of a technology is knowing exactly what to ask for and knowing it exists in the black world.) With unacknowledged energy technology, civilian physics research is being forced to re-invent fueless energy sources slowly and painfully, such as zero-point energy extraction. Bennett Hart, Deputy Director of the National Reconnaissance Organization (NRO is bigger than the CIA) explained to me that many of the exotic devices with high classification levels often move up in secrecy levels until they are "out of sight." He went on to say, "It is easier for us to pay a private contractor to re-invent something so it will come out at a lower classification level, than to try to declassify it." He told me, at the International Space Development 2002 conference in DC, he would attempt to investigate the existence of **inertia-free shielding** (after I explained to him what it was and how useful it is) and do what he could on its declassification but I'm not holding my breath.

Regarding the existing conundrum of interplanetary travel, with our present lack of appropriate propulsion technology and cosmic ray bombardment protection, we mentioned in the last chapter that Arthur C. Clarke has predicted, that in 3001 the "inertialess drive" will most likely be put to use like a controllable gravity field, thanks to the landmark paper by Haisch et al.[187] This will have to be a requirement of interplanetary travel in the near future. As the reader now knows from studying the discovery of Haisch, Rueda and Puthoff on inertia in the previous chapters, there is an electromagnetic formula awaiting discovery that will interrupt the natural interaction of a body with the surrounding electromagnetic ZPE sea, in the shorter wavelength (high frequency) region for this to operate. Safe space travel does not seem so remote anymore.

Figure 8.1 Quantum Effect on Reality
By Ruben Bolling TomDBug@aol.com
Physics Today, 1997

Chapter 8

Zero Point Field Consciousness

Overview

The Figure 8.1 cartoon offers a humorous and inspiring starting point for this chapter which asks the question, "Is there any connection between zero-point energy and consciousness?" Ever since the early days of quantum theory a hundred years ago, the **wave-particle duality** has perplexed almost everyone who learns about it. Without going into a lot of detail, since it is described everywhere in the physics literature, let me just point out that the images in Fig. 1 and Fig. 2 of the cartoon are the basis of the duality. Furthermore, the experiment described can be done equally well with laser light, electrons, or even neutrons! There is great significance to the particle's intermediate state of "wave function," "wavicle," "matter wave packet," "probability function" or photon as <u>it literally goes through both slits of the double slit filter</u> and, as the great Paul Dirac eloquently explained, "*interferes with itself*." Instead of regarding it as "either-or" situation, quantum physicists are forced to admit the amazing ability of a single particle to extend throughout space until it is called upon to *collapse the wave function* and limit its size to a spherical object as it hits something, "fixing its location." (It is well known in undergraduate physics that the equation describing **the wave function extends to infinity**, thus explaining the enormous size of each matter wave.)

When we humans see the macroscopic separation distance of the two slits compared to the classical femtoscopic size of the electron, it inspires many physicists, such as Werner Heisenberg, to become philosophical about the nature of reality.[188] Furthermore, there is an analogy of a primitive form of consciousness or awareness in the simple "connectedness" of distant physical matter that makes the field more fundamental than matter or even forces. This argument has its roots in such experiments and the underlying quantum physics theory. Heisenberg puts it quite simply,

> For instance, the great scientific contribution in theoretical physics that has come from Japan since the last war may be an indication for a certain relationship between philosophical ideas of the Far East and the philosophical substance of quantum theory. It may be easier to adapt oneself to the quantum-theoretical concept of reality when one has not gone through thee naïve materialistic way of thinking that still prevailed in Europe in the first decades of this century.[189]

It is interesting that the materialistic way of thinking Heisenberg mentions still prevails today in Western science. However, quantum theory has opened up the opposite of matter (the field of consciousness) to scientific analysis.

There is a book called *The Field* that several people told me to read. When I finally scanned through it, I started to rethink the connection that consciousness might have to the ZPF and then remembered its scientific basis in quantum mechanics. Especially when considering *non-local* effects and *entanglement*, it is understood that the ZPF is the medium that such actions at a distance operate through. Also intriguing is the fact that ZPF is technically analogous to the traditional characteristics of some aspects of God, as my quantum mechanics professor, Dr. Y.C. Lee suggested to his class years ago. Therefore, we seem to have common ground for bridging consciousness and matter, mind and the ZPF. That is what author Lynne McTaggart, Dr. Fred Alan Wolf,[190] Dr. Kenneth Pelletier[191] and others believe as well. Quantum physics and the ZPF also force the scientist toward something beyond determinism as well.

Figure 8.1 illustrates the observer's contribution to quantum measurements. Starting with Einstein and Bohr and up until modern times, for example, the problem of *quantum wave functions* has challenged physicists to recognize that observation or measurement necessarily interferes with reality. The situation quickly evolved to the point where many quantum physics books also emphasize the role that consciousness has within quantum reality.[192] Since the quantum field they are talking about is the quantum vacuum, their conclusions also affect how we regard the larger implications of the role that the ZPF plays in our daily lives. This book would be incomplete without examining the perplexing questions and deeper thoughts about something as omnipresent as the ZPF, which is full of enormous energy density, interacting and connecting intimately with every part of physical reality.

The Non-Local Universe

Just as the McTaggart book presents a larger picture of the "connectedness" that an all-pervading zero-point field can include, so does a book by Drs. Robert Nadeau and Menas Kafatos called *The Non-Local Universe*.[193] They point out that quantum mechanics experiments by Nicolus Gisin at the University of Geneva have now proven "**entanglement**" for particles up to seven miles apart. This is a more modern term for the connectedness that Einstein, Podolsky and Rosen (EPR) warned about with a specific experiment of connected particles, thinking that objective reality can contradict the uncertainty principle, as Neils Bohr replied by insisting that everything in quantum mechanics is random though a complimentary connectedness exists: if you know something about one entangled particle, you instantly create the complimentary quantity with the other.[194] Gisin calls this entanglement "dramatic evidence that nonlocality is a fact of nature."

The "signals" did not weaken or diminish over the large distance. Nadeau and Kafatos point out that if they had weakened or diminished, "physical reality would be local in the sense that nonlocality does not apply to the entire universe." This did not prove to be the case. "And this obliged the physicists to conclude that **nonlocality or non-separability is a global or universal dynamic of the life of the cosmos**."[195] The proper way to view the correlations he concludes, is that they occurred in "no time" in spite of the vast distances involved between the detectors. The reason for this is that experiments with several meters of separation by Alan Aspect at the University of Paris years before proved the effect travels much faster than the speed of light. Of course, more importantly, the distance traversed in space is the ZPF, which pervades all of the universe. Brian Greene, author of *The Fabric of the Cosmos*, says about this famous experiment, "*Entangled particles, even though spatially separate, do not operate autonomously.*"[196] Then, the question that remains is how many entangled particles exist in the universe or what percentage of particles are entangled?

The authors of *The Non-Local Universe* point out that the quantum vacuum has had billions of years to evolve and in that amount of time, it is likely that every particle in it has at some time become entangled with each other. From our newly gained knowledge of ZPE, we can add to that the entanglement of virtual particles, the connectedness of everything and everybody in the universe with every other body becomes a physics principle instead of an esoteric belief.

For those looking to connect physics and consciousness, this can be a rich starting place for fruitful discussions and even theoretical calculations, as Nadeau and Kafatos suggest.

Consciousness and the Zero Point Field

There are some intriguing books that dive deep into this topic and provide a wealth of information to meditate on. For example, *Mind Underlies Spacetime* makes the case for the philosopher's, ideal of one Mind being more fundamental than matter with a good dose of Kant, Spinoza and a few physicists' description of geometrodynamics and general relativity.[197] Another text, *The Hidden Domain*, makes the case for the electron having consciousness, quoting the physicist David Bohm (who is famous for his theory of the "implicate order"):

> Maybe it is! What sense would it make to say it is not? The electron must behave in all sorts of strange ways, like being

> a wave and a particles at the same time and jumping from one state to another without passing in between – and doing all sorts of things that cannot be understood but only calculated. If you don't want to say it's alive I suggest that you should say that the electron is a total mystery and all you can do is to compute statistically how it will reveal itself phenomenally in certain kinds of measurements.[198]

This is a provocative book for having chapters like "Consciousness as Energy" and "Consciousness and the Wave Function" which reminds me of a colleague years ago who wrote a paper with a wave function analysis of a person. However, to answer the question posed at the beginning of this section, the author concludes the wave function topic with a good summary from David Bohm, who compares mind and matter to two complimentary quantities in quantum mechanics:

> That is, if consciousness is a separate phenomenon, then matter and mind are separate, and a theory is required for their interaction. A more promising approach is the transcendental point of view of Bohm: that is, matter and mind are aspects of one overall reality. Since consciousness, or mind, is not directly accessible to our physical instruments, then matter is the form that consciousness takes when introduced or projected into our everyday world.[199]

New Scientist reviews a new book entitled, *Quantum Enigma: Physics Encounters Consciousness* but criticizes the book for not spelling out where the connection between the quantum world and consciousness is.[200] It seems that the ZPF provides the medium for such an exchange, according to Dr. Jordan Maclay.[201]

Subquantum Kinetics

A colleague of mine, Dr. Paul LaViolette, has written a book about what it happening in the negative energy levels below the phenomenal ZPE. It is an interesting and unique book that also predicts electrogravitics and other experimental outcomes of his theory of subquantum activity of the vacuum.

LaViolette asks the question, "Where does the zero-point fluctuation activity gets its energy?" His answer is the following:

> Subquantum kinetics provides a ready answer. The zero-point energy is ultimately animated by the etheric Force, the Prime Mover that impels the reaction and diffusion activity of its sea of etherons. It is the randomly varying concentrations of these animated subtle etheric particles that constitute the spontaneously arising zero-point fluctuations. We should not use the term 'energy' to refer to this animating source. Whereas field potential is an observable quantity, etheric Force is not. Like the ether itself, its existence must be inferred.[202]

His physics principle of an unknowable etheric Force thus offers another perspective on the quest to understand a more fundamental basis for the ZPF.

The Synchronized Universe

An different perspective comes from the Statistical Electrodynamics (SED) group that proposes a universe filled with photons (about 10^{80} of them to be precise) that create a random distribution which gives rise to the unpredictability of quantum mechanics. It is a classical approach to ZPE without quantum mechanical rules. While acknowledging Boyer, Puthoff, Haisch and Rueda, one physicist, Dr. Claude Swanson, also says that "consciousness must affect the Zero-Point energy in the vacuum."[203] Swanson describes a theory of how photons can become synchronized and attempts to explain psychokinetic (PK) experiments with the theory and proposes that the **background zero point energy can be strongly affected by the mind** as it focuses these photons.

It is worthwhile mentioning the other supporting work in the area, such as many Qigong masters, some of whom have demonstrated their abilities at the National Institute of Health here in Washington DC.[204] I'm also familiar with the work of Jack Houck, who I have seen demonstrate his skills several times and also teach a crowd to replicate similar results. He calls the effect of the mind on metal, "warm-forming" since the metal bar gets warm to the touch. Jack also has electron microscope pictures of the disruptive effect a successful bend will cause (see www.JackHouck.com for more information). Lastly, I'm also acquainted with Cleve Backster, who I have seen a few times. I've also written a paper about the protocol that Cleve follows to take samples of the oral leukocytes from the mouth and apply a gold electrode to them in order to watch their electrical activity.[205] A book has been written about the extraordinary connectedness that biological cells can demonstrate with their distant host, even if miles apart.[206]

Summary

Of course, all of this material has to end up with a mystical conclusion. While Sir James Jeans is famous for stating that the universe looks more like a great thought than a great machine in the early part of the last century, there are also spiritual leaders who offer more personal perceptions. If you tend to be religious, then this will appeal to you. If not, please accept my apologies for diverging a bit in my quest for the truth underlying ZPE:

> If you saw God right now, you would see Him as one mass of light scintillating over the whole universe. As I close my eyes in ecstasy everything melts into that great Light. It is not imagination; rather, the perception of the Sole Reality of being. Whatever is seen in that state will happen; that is the proof of the reality of the Omnipresent Light of all becomings.[207]

This is the most inspiring passage that I could finish this chapter with, offering a glimpse of what perhaps is behind all of those photons of light in the zero point field. It is also notable that quantum physicist, Bernard Haisch, who has contributed so much to ZPE theory, has a book entitled, *The God Theory*, which has received great reviews.[208]

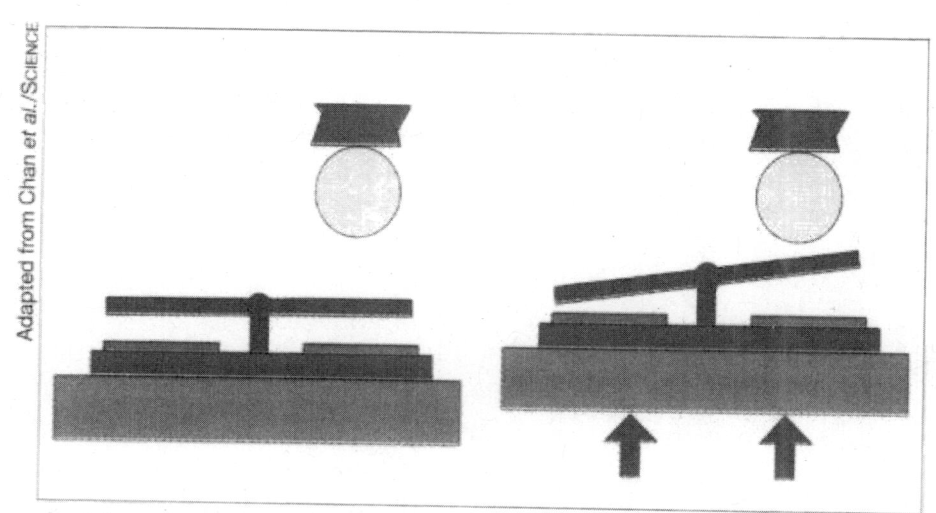

A chip with a see-saw plate suspended parallel to its surface (left) is pushed up (right) toward a ball. Quantum fluctuations in empty space produce a force that tilts the plate.

Figure 9.1
Bell Labs, Murray Hill, NJ
Courtesy of *Science News*

Chapter 9

Zero Point Energy Effects & Magic Tricks

Overview

There are a lot of places that ZPE can be seen at work. For a ZPE "fuel" which you can use in the near future, we need to review some of the magic that has been discovered in nanotechnology, just in the past few years. Ever since the National Nanotechnology Initiative, the effects of ZPE have been showing up in a number of places. Most of all, these effects involve the Casimir force, which in Fig. 9.1 for example, cause a seesaw balance beam to tilt toward the ball just based on its proximity to the object.[209] This action alone utilizes ZPE since work is being done without an expenditure of our outside input energy. The ZPE Casimir force F is exerted over a distance x causing the beam to bend, thus performing useful work (but subtracting the energy required to move the ball into position). However, even more unusual and potentially useful effects have manifested in laboratories around the world. This chapter will show you some of the unusual tricks that ZPE can do.

Casimir Seesaw

The experiment that is shown in Fig. 9.1 has been built by Professor Decca and offers a glimpse into the magic that the Casimir force can cause on a micro scale. In Fig. 9.2, we see a more detailed image of the same setup which is also shown in its micro reality in Fig. 9.3, where the plate is much larger than expected from the drawing, in order to maximize the force being applied. Whether this device can produce an oscillatory effect is open to discussion. If so, it would open the possibility of charge transfer and perhaps work done by the Casimir force.

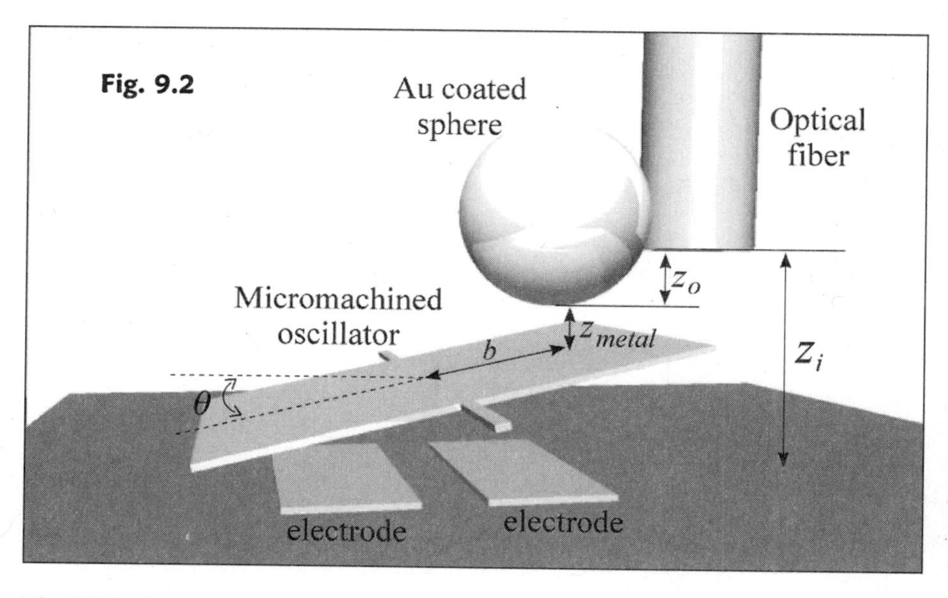

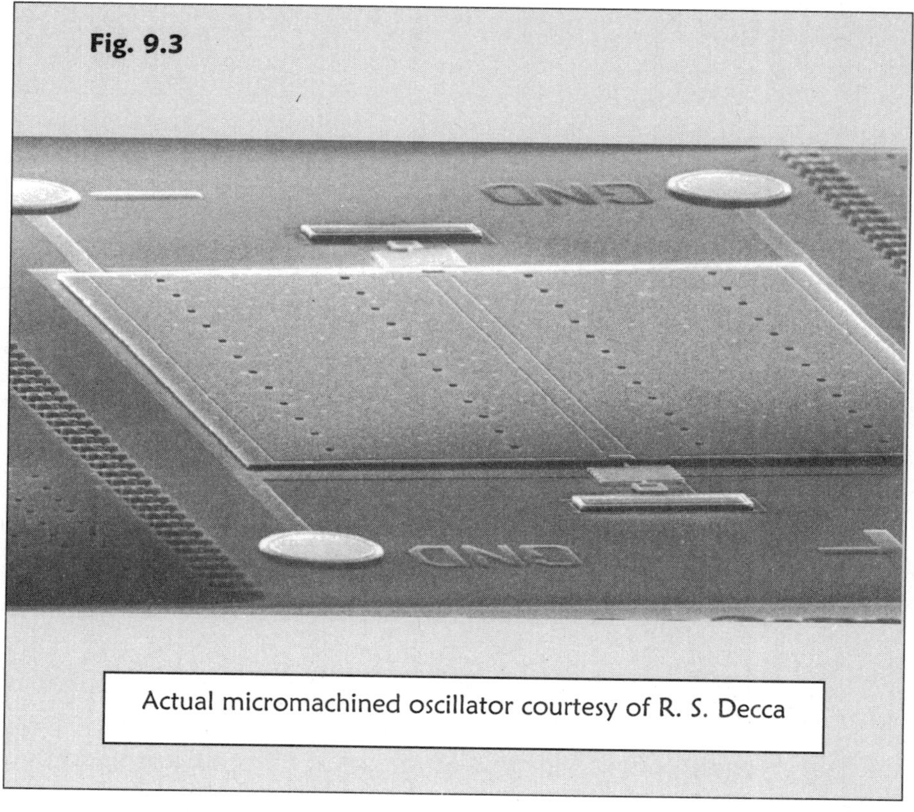

Actual micromachined oscillator courtesy of R. S. Decca

Notice that the actual photo of the cantilever (Fig. 9.3) appears to be so close to the electrodes that it seems to be touching them. That is because the distance of separation is less than a millionth of a meter (micron).[210]

Zero Point Energy Effects

A first important step is to look at *all of the effects that exhibit ZPE.* These are regarded by the experts as sufficient "proof" of the existence of ZPE. Feel free to copy and use the list on the next skeptic you meet. Remember that *ZPE evolves from both classical electromagnetism (E&M) and from quantum mechanics.* Therefore, if confusion or disbelief sets in, simply recall that E&M treats everything in nature as waves. Virtual particles can therefore be regarded fundamentally as ephemeral E&M waves or pulses of light. For example, Dr. Frank Mead (US Patent #5,590,031 seen in Fig. 3.1) calls ZPE "zero point electromagnetic radiation energy" following the tradition of Timothy Boyer who simply added a randomizing parameter to classical ZPE theory, thus also inventing "stochastic electrodynamics" (SED), which is a only a randomized E&M theory.[211] Lamoreaux, on the other hand, refers to it as "a flux of virtual particles", because the particles that react and create some of this energy are popping out of the vacuum and going back in.[212] The *New York Times* simply calls it "quantum foam." But the important part about ZPE comes from Dr. Robert Forward who states emphatically, **"the quantum mechanical zero point oscillations are real."**[213]

There are some people today, I have found, who still say they "don't believe" in ZPE (even key physicists at the National Science Foundation and other branches of the government), as if a *religious attitude* was required. A more rational approach would be to look at the scientific literature, which provides an overwhelming wealth of experimental evidence for the existence of ZPE, including the following list below, that grows larger each month (note references for each entry):

1) Anomalous magnetic moment of the electron[214]
2) Casimir effect[215]
3) Diamagnetism[216]
4) Einstein's fluctuation formula[217]
5) Gravity[218]
6) Ground state of the hydrogen atom[219]
7) Inertia[220]
8) Lamb shift[221]
9) Liquid Helium to T = 0 K[222]
10) Planck's blackbody radiation equation[223]
11) Quantum noise[224]
12) Sonoluminescence[225]
13) Spontaneous emission[226]
14) Uncertainty principle[227]
15) Van der Waals forces[228]

When each of these references for the corresponding phenomenon are examined closely, it becomes hard to avoid the fact that the argument for the reality of zero point energy carries substantial weight. This necessarily includes the enormous energy density associated with ZPE as well. A further argument, which builds a case for the fascinating implications of ZPE as the "fuel of the future" by extracting some of the ZPE energy for useful work, is made in subsequent chapters.

In my feasibility study called, "***Practical Conversion of Zero-Point Energy***," a list of techniques or "tricks" are included in the back entitled, "Vacuum Engineer's Toolkit." This is a valuable and indispensable toolbox for scientists and nano-engineers who wish to know, "what can we use to manipulate ZPE for our advantage?" In this chapter, a sample of these amazing and unique tools are explained in enough detail to gain your interest and appreciation. It is important to realize that the quantum world of ZPE has a wealth of properties that are *totally different* than the macroscopic world of Newtonian physics, because much of quantum mechanics is predominantly counter-intuitive. I believe that the ZPE properties or tricks are the means that God has offered us to manipulate ZPE to perform useful work. Knowing the tricks of the trade is always necessary to succeed in any field.

Focusing Vacuum Fluctuations

Figure 9.4
Electromagnetic Wave

For our first magic trick, we start off with a development that directly concentrates ZPE electromagnetic energy to a point, called "focusing vacuum fluctuations." Much like concentrating solar energy causing a great increase in temperature at the focal point, utilizing a parabolic mirror in both cases accomplishes the same thing. The mirror designed for this purpose needs to be about 1 micron in size (Figure 9.5). Considering the "plasma wavelength" in the range of 0.1 micron for most metals, Ford predicts that it may be possible to deflect atoms at room temperatures and <u>levitate them</u> in a gravitational field, and even trap them within a few microns of the focal point F.[229]

The reason he can do this is that depending upon the design parameters, you can get a <u>repulsive force</u> at the focus with a region of negative energy density, or an <u>attractive force</u> at the focus with a region of positive energy density. Note: **a positive energy density results in an attractive force.** It also has a great benefit in that it is a totally passive device. In other words, *this type of trapping would require <u>no</u> externally applied electromagnetic fields or photons or electrical energy input.* It represents a passive collector.

The enhanced vacuum fluctuations responsible for these effects are found to arise from an interference term between different reflected rays. The interesting conundrum is the suggestion that parabolic mirrors can focus something even in the absence of incoming light. However, we need to keep in mind that vacuum fluctuations are often treated as evanescent electromagnetic fields. The manifestation of the focusing phenomenon is the growth in the energy density and the mean squared electric field near the focus.[230]

Focusing vacuum fluctuations in many ways resembles a similar phenomenon called "**amplified spontaneous emission**" (ASE), mentioned in Chapter 4, which occurs in a gain medium, where the buildup of intensity depends upon the quantum noise associated with the vacuum field.[231]

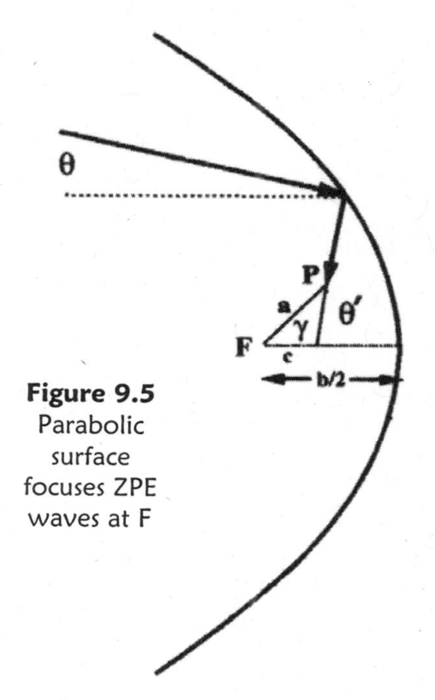

Figure 9.5 Parabolic surface focuses ZPE waves at F

The exciting part of this vacuum engineer's trick is that it can work along with a number of other tricks to enhance the efficacy of the overall design. In other words, the one with the most tricks up his sleeve wins! The classic "traveling electromagnetic wave" model of light (Figure 9.4), reprinted from Chapter 3, is used here since virtual photons are regarded as the same as light (for a very short period). This is in keeping with the laws of quantum mechanics and what is called the "uncertainty principle."

Spatial Squeezing of the Vacuum

There is nothing like the common every day experience of squeezing a tube of toothpaste to get the contents to spill out on the toothbrush, except maybe the squeezing of a shampoo bottle to wash your hair, or the tube of ointment on a cut. All of these experiences show that *force* is necessary to cause the stuff inside to spill out. Spatial squeezing (though it may seem to be a really weird magic trick) is similar to other types of squeezing. Similarly, *it decreases the energy density on one side of a surface*, below its vacuum value, *in order to increase the Casimir force*. Sort of like a trade-off for conservation of energy considerations.

For a moving surface that goes back and forth, like Drs. Pinto's or Decca's cantilevers, this can also create a correlated "excitation of frequency modes" into squeezed states and "sub-Casimir regions" where the vacuum develops structure. **"Pressing zero-point energy out of a spatial region can be used to temporarily increase the Casimir force."**[232] This spatial squeezing technique is gaining increasing acceptance in the physics literature as a method for bending quantum rules while gaining a short-term benefit, such as modulating the quantum fluctuations of atomic displacements below the zero-point quantum noise level of coherent phonon (vibrational) states.[233]

This squeezing technique involves minimizing the "expectation value" of the energy in a prescribed region, such as a cavity or an empty box. In general, a squeezed state is obtained from a particular state of what is referred to as the "annihilation operator" (somewhat like a *Nano-Terminator*) by applying to it what is called the unitary squeezing (or *dilation*) operator.[234] Ideally, **it seems promising to generate squeezed modes inside a cavity by an instant change of length of the cavity.**[235] The high speed (and frequency) required for a movable membrane might be physically challenging however. The squeezing will cause a modification of the Casimir force so that it can become a time dependent oscillation from a maximum-to-minimum force. Theorists predict that pursuing what is called "resonance measurements" may turn out to be the most realistic experimental approach in order to create a periodic variation in the Casimir force by squeezing.

The benefit of squeezing is that **the resulting emission of photons is almost double** of that allowed by the Planck radiation law, which demands quantized energy (like little packets of candy) which are also called "quantized field modes." Dr. Hu has found that the other field modes go to a strange, *mixed quantum state* due to the intermode interaction caused by the classical Doppler effect from the moving mirrors. The theory also predicts that the significant features of the nonstationary Casimir effect are not sensitive to temperature.[236] This property alone can help create stability in future ZPE generators in any hostile environment.

Boxed in Cavity

Is it possible that the **dimensions of any box** in the nanoworld is critical to whether it *helps or hurts* the energy effort? Is it possible that certain boxes give one result while other boxes of different sizes give the opposite result? Yes! It's like a "rabbit out of the hat" trick. In fact, when designing the microbox for a cantilever chamber like Pinto's or any other micron-sized ZPE converter, the size and shape is very important to helping or hurting the effort or simply pushing when you want to be pulling. To explain, in Fig.9.6 there is the **black area** and the shaded area between zero energy density and the maximum energy density. What this region corresponds to is quite simple: it is the actual dimensions of a box (in microns) that will exhibit only a positive or *attractive* Casimir force (**inward force**). This is to be expected since (see Ch. 10) the Casimir force inversely depends upon the fourth power of the distance between the surfaces. The surprising region, to those of

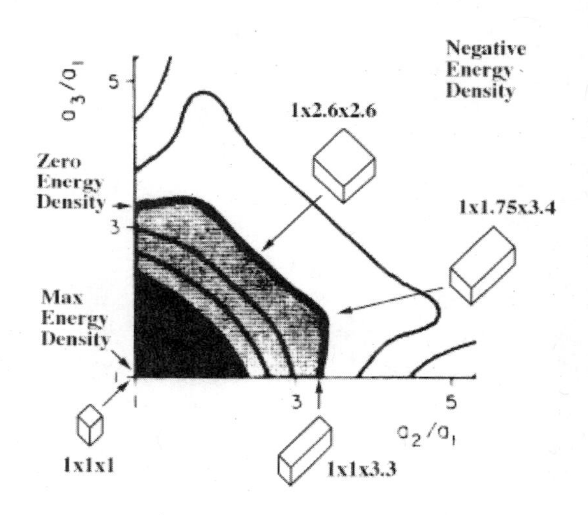

Figure 9.6
Constant positive, negative and zero Casimir force regions for various rectangular metal microboxes

us who believe Casimir might behave like the gravitational force, is that the force can also be negative or **repulsive**, as in the **white region** in Fig. 9.6.

Jordan Maclay, who received a NASA grant to pursue this work, has shown that the length, width and height of a micron-sized, conductive-metal box cavity can be graphed to provide a guide to a particular experimental design. This graph estimates the resulting Casimir force that will be present in each case. Half of the graphed region creates a **positive Casimir force** that attracts the surfaces together, while certain other dimensional choices result in a box that has a **negative Casimir force**, pushing the surfaces away from each other. This "push-pull" ability can be used to our advantage when nano-engineers are designing membrane or cantilever-sided boxes, so that the Casimir force will add energy and range to the operation of the movable surface. It is like a built-in amplifier, right inside the fabric of the universe.

The discovery by Maclay of a particular box dimension (1 x 1 x 1.7, where 1 is a micron or less), that sits in the middle of attractive and repulsive Casimir forces (Fig. 9.6), presents a possible scenario for vacuum energy extraction.

> This interesting motion suggests that we may be organizing the random fluctuation of the EM field in such a way that changes in pressure directly result, which could lead to work being done. One interesting question is can we **design a cavity that will just oscillate by itself in a vacuum.** One approach to this would require a set of cavity dimensions such that the force on a particular side is zero, but if the side is moved inward, a restoring force would be created that would tend to push it outward, and vice versa. Hence a condition for oscillation would be obtained. Ideally, one would try to choose a mechanical resonance condition that would match the vacuum force resonance frequency. More complex patterns of oscillation might be possible. The cavity resonator might be used to convert vacuum fluctuation energy into kinetic energy or thermal energy. More calculations of forces within cavities are needed to determine if this is possible, what would be a suitable geometry and how the energy balance would be obtained.[237]

Maclay concedes however, that upon analyzing Forward's charged parking ramp, with like charges supplying the restorative force to the Casimir attractive force, that no net work would be done for any given oscillation cycle.

A great magic trick of applying Maclay's analysis, the Casimir force that comes into play here is also something that biology already deals with on a daily basis. At the Second Conference on Future Energy, hosted by our institute, Dr. Pinto explained that a certain South American gecko has pads on its fingers which are so smooth that the Casimir force has been found to be the only explanation for why it can adhere to any surface. Another example of the Casimir force in biology is certain spores and bacteria which are a micron in size. It has been calculated that they have to exert an extra tension to keep their bodies from collapsing under the Casimir force. For a dilute, dispersive dielectric ball for example, the Casimir surface force is found to be attractive with inward pressure.[238] One application for this type of Casimir force calculation lies with biological cells which are spheres with a high dielectric constant. Figure 9.7 shows a B-lymphocyte which is 1 micron across which therefore must experience and compensate for the inward Casimir pressure. Dr. Jordan Maclay explains,

> Biological structures may also interact with the vacuum field. It seems possible that cells, and components of cells, for example, the endoplasmic reticulum may interact with the vacuum field in specific ways. A cell membrane, with a controllable ionic permeability, might change shape in such a way that vacuum energy is transferred. Microtubules, in cell cytoskeletons, may have certain specific properties with regard to the vacuum field. Diatoms, with their ornate geometrical structures, must create interesting vacuum field densities; one wonders if there is a function for such fields."[239]

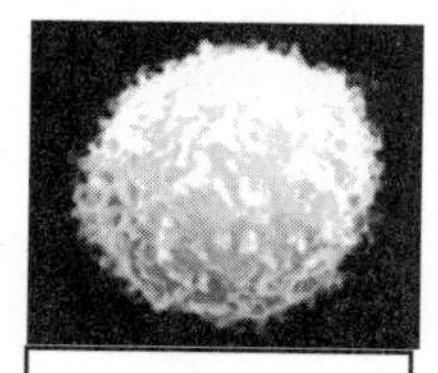

Fig. 9.7 B-Lymphocyte

Many of these structures that are less than one micron in size have much higher Casimir pressures to contend with, such as ribosomes which are about 0.02 micron across.[240]

Forces That Bind

Anyone working in the realm of a micrometer (a millionth of a meter, also called a "micron") to a nanometer (a

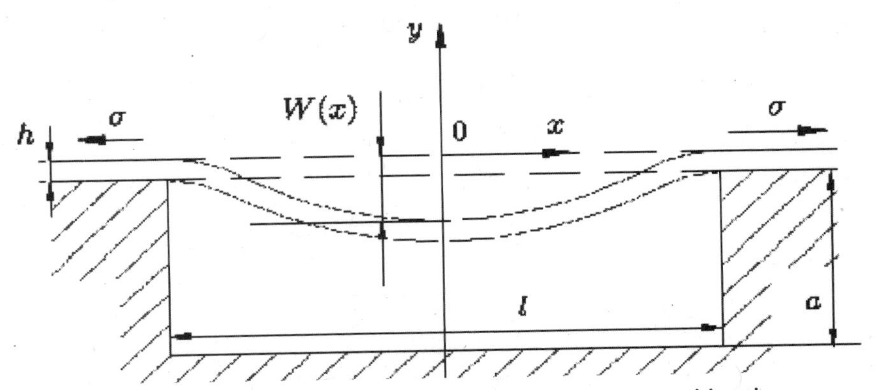

Figure 9.8 - NEMS cantilever bridge deflection caused by the Casimir force

billionth of a meter or 10^{-9} m) will confirm that allowing two parts of their favorite cantilever or movable part (as in Fig. 9.1) to touch will usually be fatal for the device. That is because the Casimir force becomes extremely strong as the small parts come together and contact each other, increasing by the third or fourth power of the separation distance. Normally, the standard complaint is "they cannot be separated" or "they can only be destructively separated." All of a sudden, the creative vacuum engineer might say, "Who needs Elmer's glue anymore?"

In Fig. 9.8 we see an close-up example of a tiny box with one movable membrane on top that is being deflected by the Casimir force from the other side, since it is less than one micron away. There are equations that are reviewed in my *Feasibility Study* regarding this invention which allow the nano engineer to solve for any desired deflection.[241] However, since the Casimir force in these equations increases by the fourth power of the distance (the difference between 'a' and W(x)), it is also regarded as a positive feedback system, with a tendency of increasing any deflection in a direction toward structural failure. (Stable systems in nature always have negative feedback.) Therefore in this situation, not only do we have proof of the

Casimir force effect on the micro scale but with such unstable positive feedback, the danger is that any small deflection may easily be amplified until "stiction" (a technical word for stickiness) causes it to fail.

Heat from ZPE

In what may seem to appear as a major contradiction, it has been proposed that, at least in principle, basic thermodynamics allows for the extraction of heat energy from the zero-point field via the Casimir force. "However, the contradiction becomes resolved upon recognizing that two different types of thermodynamic operations are being discussed."[242] Normal "thermodynamically reversible" heat generation process is usually limited to temperatures above absolute zero ($T > 0$ K). "For heat to be generated at $T = 0$ K, an irreversible thermodynamic operation needs to occur, such as by taking the systems out of mechanical equilibrium."[243] The authors give examples of systems with two opposite charges in a perfectly reflecting box being forced closer and farther apart. With adiabatic expansion and contraction curves, they propose that heat is extractable in principle. Though a practical method of energy or heat extraction is not addressed in the article, the theoretical basis for designing one is given a more firm physical foundation.

Shape and Depth of the Casimir Force

Some of the more recent developments in the exploration of the Casimir force are shown in the last few figures. For example, in Fig. 9.9, Dr. Budker from UC Berkeley shows that the shape of the probe tip will affect the intensity of the force. The magic here is the "diffraction" concept for virtual photons which is different for the two probe designs.[244] A new aspect of the Casimir force has been discovered by Dr. Iannuzzi by coating his probe with a mirror (thick film) and comparing it with a transparent (thin film) coating (Fig. 9.10).

Effects of Edges

shape of decay function is strongly dependent on size and separation of surfaces

REF http://images.google.com/imgres?imgurl=http://www.sr.bham.ac.uk/yr4pasr/project/casimir/currentthumb.jpg&imgrefurl=http://www.sr.bham.ac.uk/yr4pasr/project/casimir/&h=275&w=275&sz=41&tbnid=Buy2QDUNZEvi6M:&tbnh=109&tbnw=109&hl=en&start=20&prev=/images%3Fq%3Dcasimir%2Beffect%26svnum%3D10%26hl%3Den%26lr%3D%26sa%3DG

- Dist > 25μm: dome shape
- The Casimir force occurs when virtual photons are restricted
- The force is reduced where virtual photons are diffracted into the gap between the plates
- Unshaded areas correspond to higher Casimir forces
- Casimir force is decreased at the edges of the plates

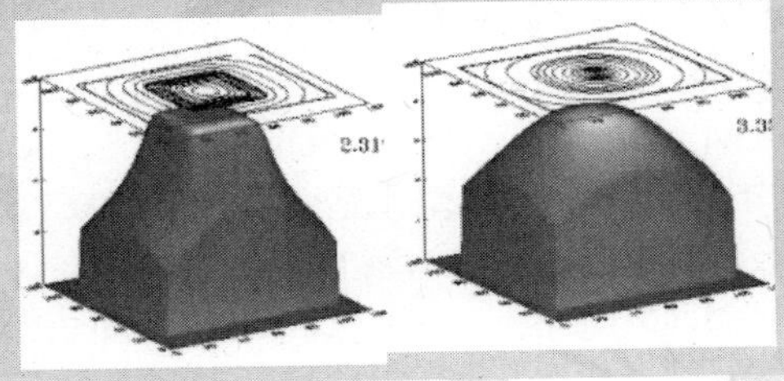

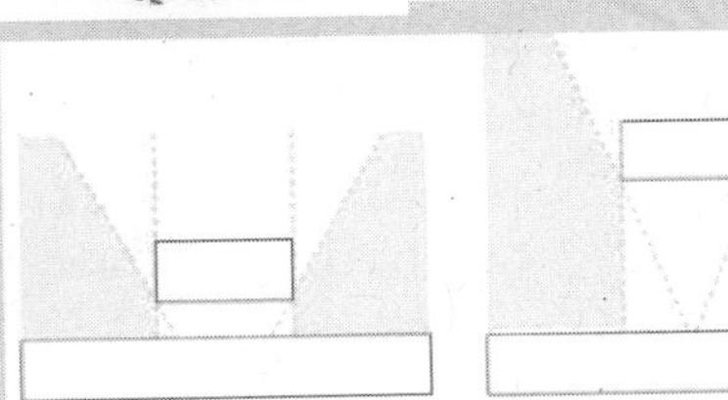

Dr. Budker, U. Berkeley

Eric Corsini

Fig. 9.9

Casimir force: skin-depth effect

When is a metallic layer really a mirror?

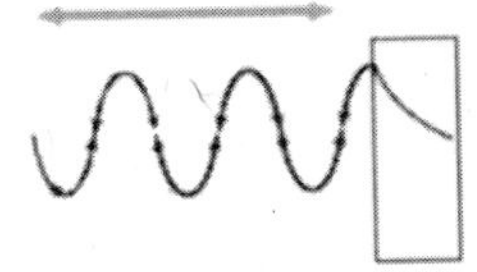

Thick (bulk-like) metallic film: reflective

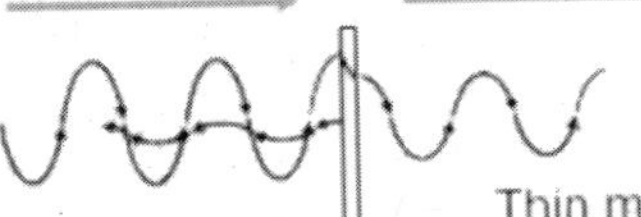

Thin metallic film: transparent (but still metallic!)

Transparent sphere coated with

Thin metallic layer

Thick metallic layer

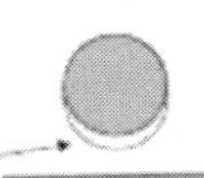

davide iannuzzi - vrije universiteit amsterdam

Fig. 9.10

The well-known "skin depth effect" is normally seen where radio frequencies go into a surface to a certain depth before they attenuate to nothing. The magic trick here is that the Casimir force, consisting of lots of frequencies, also is proven to have skin depth, as seen in Fig. 9.11. For any given separation distance, like 0.2 micron, the thick reflective film in both experiments helps create a larger force (about 25 piconewtons) than the transparent film. One way to understand this is that the Casimir force of attraction works by excluding electromagnetic waves from the cavity, which is what a mirror is good for.

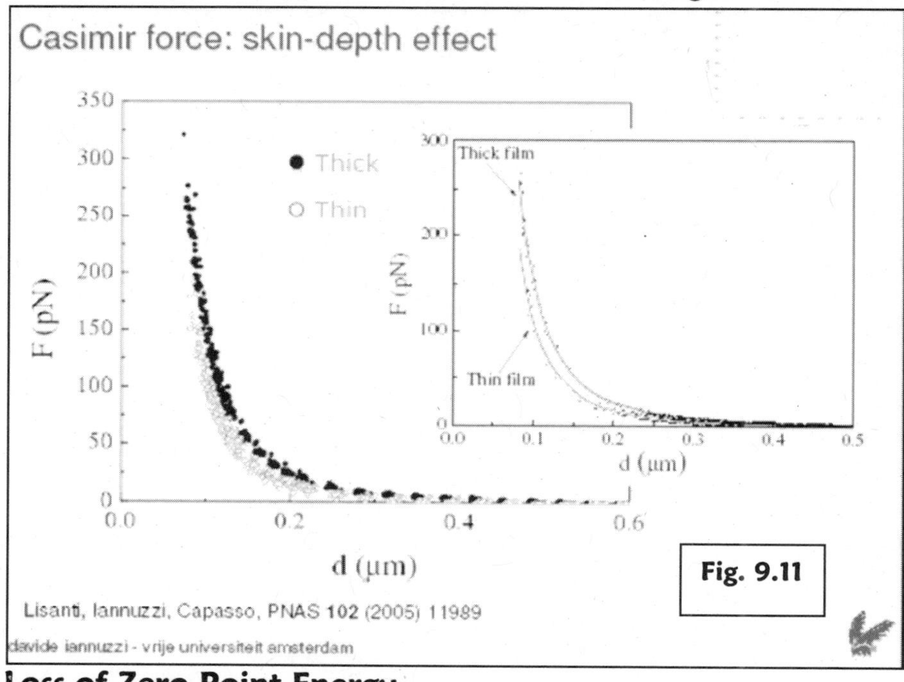

Fig. 9.11

Loss of Zero Point Energy

In organic chemistry, the ZPE (lowest vibrational energy state) of a molecule is relatively large. For C—H bonds in alkanes is about 4 kcal/mole. Furthermore, what may seem to be magic is the loss of ZPE in a change from reactant to transition state for the C—H bond. This will also affect the reaction rates and activation energy, proportional to how much ZPE is converted.[245] In other words, nature is using and converting ZPE on a regular basis during chemical reactions.

Wormhole from ZPE

We finish this chapter with what is certainly the most unusual magic trick with ZPE effects, Dr. A. Popov has published self-consistent solutions of the Einstein field equations that describe a "long throat of a traversable Lorentzian wormhole."[246] This type of solution has been done before but in this case, Popov is using an electrostatic field and the vacuum fluctuations themselves, as a quantized fields for the source of this uniquely curved spacetime.

For Further Information

Many other examples of systems that work in apparent violation of thermodynamic laws are given in my other book, *Practical Conversion of ZPE*. May I also suggest *The Quantum Vacuum* book by Peter Milonni as an excellent reference for those looking for more magic from ZPE. For example, his explanation of how an electron "can see" the presence of the walls of a cavity which affect its resonant frequencies is priceless:

> ...how can the emission of a photon be affected by an atom's environment when the atom can only "see" its environment by emitting a photon in the first place? Such an objection is invalid. As long as the emission lifetime is large compared with 2z/c [z is distance to wall], the atom has ample opportunity to "see" its environment. But it need not emit a photon to do so.[247]

This quote tries to explain in quantum mechanical terms the magical effect of the atom's wave function already being spread out to the limits of its environment. Milonni also recommends Boyer's article on ZPE and long-range forces. Another excellent text on the scientific details is *Van der Waals Forces* by V. A. Parsegian which works out a lot of good examples.[248]

Figure 10.1 ZPE was the cause cited in *Science* magazine for the "Breakthrough of the Year" which revealed the discovery that galaxies are accelerating away from each other.

Chapter 10

The Science Behind the ZPE Effects

Overview

In this chapter, a cursory summary of the physics behind the ZPE effects is presented for the non-technical audience. A more detailed, rigorous review is to be found in my feasibility study called "*Practical Conversion of Zero-Point Energy*" It begins with a look at a British development called "NANOCASE" that supports ZPE research on a very significant budget scale. It simply looks for better measurements of the Casimir force and anything is better than no funding at all.

Nanocase - Nano-scale machines exploiting the Casimir Force

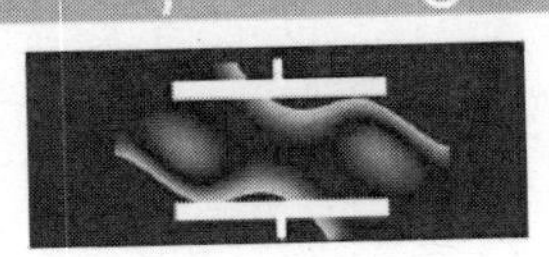

- *UK: University of Leicester*
  - *Prof. Chris Binns*
  - *Dept. of Physics and Astronomy,*
  - *E-mail: cb12@le.ac.uk UK: University of Birmingham*
- *France: Université Pierre et Marie Curie*
- *Sweden: Linköping University*

- *Duration*
  - *36 months*
- *Project costs*
  - *€ 799 991*
- *EU funding*
  - *€ 799 991*
- *Project reference*
  - *Contract No. 12142 (NEST)*
- *Web: http://www.cordis.lu/nest*

COFE 2
Sep 23 2006

Zero Point Energy Advances
Dr. Thorsten Ludwig

Figure 10.2

What Lies Beneath the Void - NANOCASE Project

Professor Chris Binns (Physics and Astronomy) at the University of Leicester has received (in 2005) European funding for an exciting project to measure the force of zero point energy (my thanks to Dr. Ludwig for this lead and Fig. 10.2):

> Three thousand years ago the Greek philosophers Leucippus and his student Demokritos proposed the concept of the atom, as a fundamental building block of materials, in order to circumvent a paradox that arises with continuous elements (such as earth fire air and water). They pointed out that if matter was really a continuum then you could cut it into smaller and smaller pieces ad infinitum and, in principle, cut it out of existence into pieces of nothing that could not then be reassembled. Thus, they reasoned, there must be a smallest piece of matter that could not be further divided the a-tomon (uncuttable) from which the word atom is derived. To complete the picture you also need a void in which the atoms move, a concept that produced fervent debate, for example, is the void a 'nothing' or a 'something' and is it a continuum or does the void itself have an uncuttable smallest unit.
>
> While the atom, the legacy of Leucippus and Demokritos, is now a familiar part of the scientific landscape, the true nature of the void remains a mystery. In classical Physics the void is a 'nothing', a simple absence of all matter and energy. Quantum theory tells a different story and states that the void is definitely a 'something'. It is a seething mass of 'virtual' particles that fleetingly appear into and then disappear from our observable universe. This activity, known as quantum fluctuations, corresponds to an intrinsic energy of the void, the 'zero-point energy', which, if the void were a continuum, would be infinite. It is generally believed that there is a smallest piece of void, which makes the zero-point energy finite but still colossal beyond the imagination. Each cubic millimetre of empty

space contains more than enough zero-point energy to create a new universe.

In a sense the actual value of the zero-point energy is not important because everything we know about is on top of it. According to quantum field theory every particle is an excitation (a wave) of an underlying field (for example the electromagnetic field) in the void and it is only the energy of the wave itself that we can detect. A useful analogy is to consider our observable universe as a mass of waves on top of an ocean, whose depth is immaterial. Our senses and all our instruments can only directly detect the waves so it seems that trying to probe whatever lies beneath, the void itself, is hopeless. Not quite so. There are subtle effects of the zero-point energy that do lead to detectable phenomena in our observable universe. An example is a force, predicted in 1948 by the Dutch physicist, Hendrik Casimir, that arises from the zero-point energy. If you place two mirrors facing each other in empty space they produce a disturbance in the quantum fluctuations that results in a pressure pushing the mirrors together. Detecting the Casimir force however is not easy as it only becomes significant if the mirrors approach to within less than 1 micrometre (about a fiftieth the width of a human hair). Producing sufficiently parallel surfaces to the precision required has had to wait for the emergence of the tools of nanotechnology to make accurate measurements of the force.

Figure 10.3
Prof. Binns with his atomic force microscope designed for measuring the Casimir force at the University of Leicester

In the last decade this has happened and a spate of measurements using atomic force microscopes has confirmed Casimir's prediction to a precision of about 5% and the zero-point energy of the void is an experimental reality. This is just the beginning however as the force has only been measured in very simple geometries such as flat parallel plates. More recent calculations show that the force is sensitive to geometry and by changing the materials and the shape of the cavity you can alter the magnitude of the Casimir force and possibly even reverse it. This would be a ground-breaking discovery as the Casimir force is a fundamental property of the void and reversing it is akin to reversing gravity. Technologically this would only have relevance at very small distances but it would revolutionise the design of micro- and nano-machines.

The srif2 and srif3 investment by the University of Leicester in the Virtual Microscopy Centre and the Nanoscale Interfaces Centre has put the University in a key position to take a lead in Casimir force measurements in novel geometries. It has led to the award of an 800,000€ grant (NANOCASE) from the European framework 6 NEST (New and Emerging Science and Technology) programme to lead a consortium from three countries (UK, France and Sweden). The programme will use the ultra-high vacuum Atomic Force Microscope installed in the Physics and Astronomy department under srif2 to make very high precision Casimir force measurements in non-simple cavities and assess the utility of the force in providing a method for contactless transmission in nano-machines.

The new instrumentation to be installed soon following the srif3 investment will enable researchers to extend the measurements to yet more complex shapes and, for the first time, to search for a way to reverse the Casimir force.

This new wave of measurements will enable an unprecedented level of probing of the void and will provide important information on new theories of gravity and with sufficient precision will even put limits on the true number of spatial dimensions. Knowing how zero-point energy varies with the shape of an enclosure may also give clues to the origin of so-called 'dark energy', discovered recently.[249]

The ebulletin story from the UK on NANOCASE and Professor Binn has the subtitle: **"exciting project connected to the 'zero-point energy' of space."** While NANOCASE is the largest ZPE grant so far, the details of the Casimir force, including its positive to negative force reversal, have already been worked out by Jordan Maclay with a NASA grant years ago (discussed elsewhere in this book). It is true that the replication and verification of previously theorized results will advance this science, even if only by a snail's pace.

Zero-Point Energy Basics

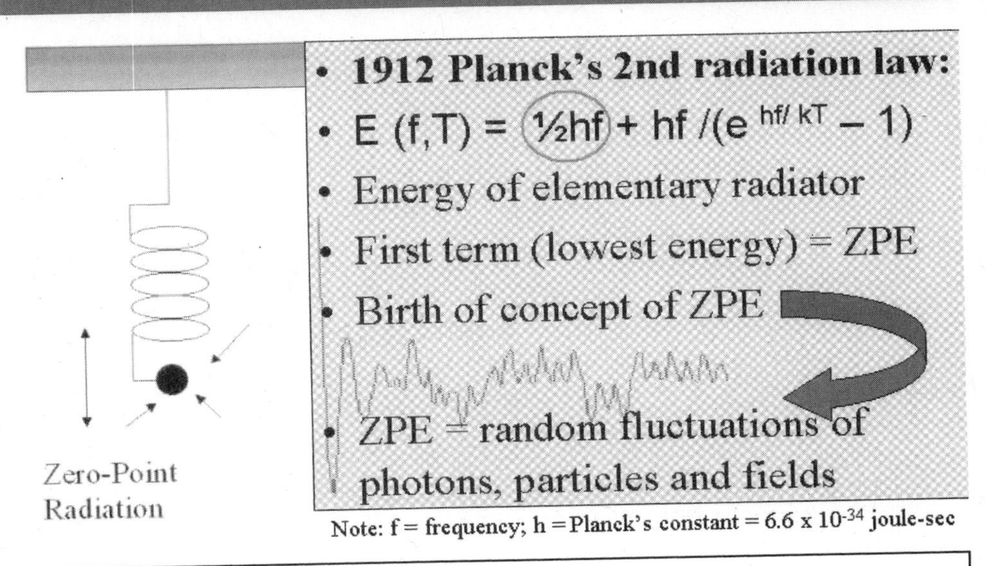

Zero-Point Energy Basics
• 1912 Planck's 2nd radiation law:
• E (f,T) = ½hf + hf /(e hf/ kT – 1)
• Energy of elementary radiator
• First term (lowest energy) = ZPE
• Birth of concept of ZPE
• ZPE = random fluctuations of photons, particles and fields
Zero-Point Radiation
Note: f = frequency; h = Planck's constant = 6.6 x 10-34 joule-sec

Fig. 10.4 Max Planck's Second Radiation Law that ushered in the modern age of zero point energy, exactly matching the "black body radiation curve"

Briefly introduced in Chapter 2 with the history of Max Planck, a Nobel Prize winner, his second radiation law made all the difference in the world. Seen in all of its glory in Fig. 10.4, the important point for non-specialists is the circled quantity **½hf**. This is the mysterious *average* of zero-point energy when looked at with quantum mechanics. I say "mysterious" because it is exactly half of what physicists normally call a quantum. For example, Planck's quantum frequency equation **E=hf** gives a clear indication of the quantum units of hf that are present in an atom (with any multiples also allowed by the same equation). However, this is nature that determined the average and it is also proven or derived from the Heisenberg "uncertainty principle" (see Appendix).

Forward to the First ZPE Converter

The first journal article publication to propose a Casimir

PHYSICAL REVIEW B VOLUME 30, NUMBER 4 15 AUGUST 1984

Extracting electrical energy from the vacuum by cohesion of charged foliated conductors

Robert L. Forward
*Hughes Research Laboratories, Malibu, California 90265**
and Air Force Rocket Propulsion Laboratory, Edwards Air Force Base, California 93523
(Received 23 November 1983; revised manuscript received 16 April 1984)

Any pair of conducting plates at close distances ($< 1\ \mu m$) experience an attractive Casimir force that is due to the electromagnetic zero-point fluctuations of the vacuum. A "vacuum-fluctuation battery" can be constructed by using the Casimir force to do work on a stack of charged conducting plates. By applying a charge of the same polarity to each conducting plate, a repulsive electrostatic force will be produced that opposes the Casimir force. If the applied electrostatic force is adjusted to be always slightly less than the Casimir force, the plates will move toward each other and the Casimir force will add energy to the electric field between the plates. The battery can be recharged by making the electrical forces slightly stronger than the Casimir force to reexpand the foliated conductor.

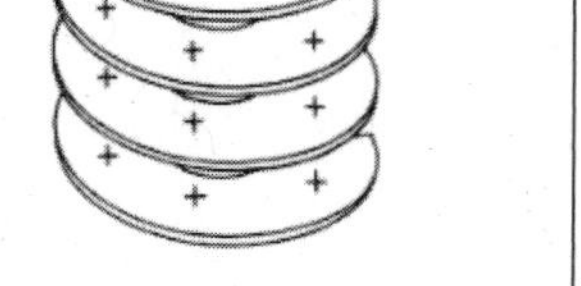

For d = 1 nm

F > 200 lb/ ft²

F > 1.5 lb/in²

ELECTROSTATIC REPULSION

VACUUM FLUCTUATION ATTRACTION

FIG. 1. Spiral design for a vacuum-fluctuation battery.

Robert L. Forward

- Casimir $F = - \pi hc / (480d^4)$
- $F = -.013 / d^4$ dynes/cm²
- Coulomb $F_{Co} = +1/8\pi\ (V^2 / d^2)$
- for d = 1 micron, $F_{Co} = F$
  …… when <u>V = 17 mV</u>
- Very little voltage is needed but really only good for electron storage battery

Fig. 10.5 Forward's force, size and voltage that he recommended machine for "the extracting of electrical energy from the vacuum by cohesion of charge foliated conductors" is summarized in Fig. 10.5.[250] Dr. Forward describes this "parking ramp" style corkscrew or spring as a ZPE battery that will tap electrical energy from the vacuum and allow charge to be stored. The spring tends to be compressed from the Casimir force but the like charge from the electrons stored will cause a repulsion force to balance the spring separation distance. It tends to compress upon dissipation and usage but expand physically with charge storage. He suggests using micro-fabricated sandwiches of ultrafine metal dielectric layers. Forward also points out (Chapter 5) that ZPE seems to have a definite potential as an energy source. Since then, it still remains an intriguing intellectual exercise as to whether stressing the Casimir force will cause any extra ZPE accentuation.

To help explore Forward's argument, we know that in general, many of the experimental journal articles refer to vacuum effects on a cavity that is created with two or more surfaces. "Cavity QED" (Quantum ElectroDynamics) is a science unto itself. For example, we also know the variation in the spacing of the spiral foils will have an effect on the quantum frequencies from the ZPF. A look into this elusive scientific world of cavity effects is found in the following quotation:

> Small cavities suppress atomic transitions; slightly larger ones, however, can enhance them. When the size of the cavity surrounding an excited atom is increased to the point where it matches the wavelength of the photon that the atom would naturally emit, vacuum-field fluctuations at that wavelength flood the cavity and become stronger than they would be in free space."[251]

It is also possible to perform the opposite feat. "Pressing zero-point energy out of a spatial region can be used to temporarily increase the Casimir force."[252] The materials used for the cavity walls are also important. It is well-known that the attractive Casimir force is obtained from highly reflective surfaces and there are more details that the "Vacuum Engineer's

Toolkit" contains[253] which indicate favorable surface effects, such as what Dr. Pinto uses in Chapter 5.

Photo-Carnot Engine

One of the main criticisms of energy extraction from the ZPF is that it represents a single low-temperature bath and the second law of thermodynamics prohibits such an energy

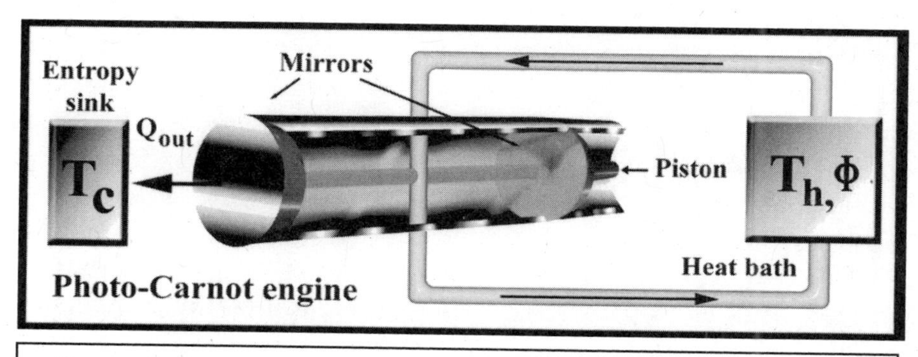

Figure 10.6 Photo-Carnot heat cycle diagram. Q_{in} is provided by hot atoms from a single heat bath.

conversion. It is well-known that Carnot showed that every heat engine has the same maximum efficiency, determined only by the high-temperature energy source and the low-temperature entropy sink. Specifically, it follows that no work can be extracted from a single heat bath when the high and low temperature baths are the same.

However, a new kind of **quantum heat engine** (QHE) powered by a special "quantum heat bath" has been proposed by Scully et al. which allows the extraction of work from a single thermal reservoir. In this heat engine, radiation pressure drives the piston and is also called a "Photo-Carnot engine." Thus, the radiation is the working fluid, which is heated by a beam of hot atoms. The atoms in the quantum heat bath are given a small bit of quantum coherence (phase adjustment) which becomes vanishingly small in the high-temperature limit that is essentially thermal. However, the phase associated with the atomic coherence, provides a new control parameter that can be varied to increase the temperature of the radiation field and to extract work from a single heat bath. The second law of thermodynamics is not violated, according to Scully et al., because the quantum Carnot engine takes more energy, with microwave input, to create the quantum coherence than is generated.[254]

The Photo-Carnot engine, shown in Figure 10.6, creates radiation pressure from a thermally excited single-mode field that can drive a piston. Atoms flow through the engine from the T_h heat bath and keep the field at a constant temperature for the isothermal 1→ 2 portion of the Carnot cycle. Upon exiting the engine, the bath atoms are cooler than when they entered and are reheated by interactions with the blackbody at T_h and "stored" in preparation for the next cycle.

The stimulus for the work came from two innovations in quantum optics: the micromaser and microlaser (Fig. 5.4) and **lasing without inversion** (LWI). In micromasers and microlasers, the radiation cavity lifetime is so long that a modest beam of excited atoms can sustain laser oscillation. In LWI, the atoms have a nearly degenerate pair of levels making up the ground state. When the lower level pair is coherently prepared, a small excited state population can yield lasing (without inversion).

Q_{in} is the energy absorbed during the isothermal expansion and Q_{out} is the energy given to the heat sink during the isothermal compression. However, instead of two states that would render this a classical engine, the QHE has three states, which can result in quantum coherence. If there is a phase difference between the two lowest atomic states, then the atoms are said to have quantum coherence. This can be created by a microwave field with a frequency that corresponds to the transition between the two lowest atomic states. *Quantum coherence changes the way the atoms interact with the cavity radiation by changing the relative strengths of emission and absorption.*[255]

In the Photo-Carnot engine, as the atoms leave the blackbody at temperature T_h they pass through a microwave cavity that causes them to become coherent with the phase before they enter the optical cavity. The efficiency of the quantum Carnot engine can exceed that of the classical engine – even when $T_c = T_h$. It can therefore extract work from a single heat bath.[256]

Inexplicably, Scully et al. fail to cite a previous work by Allahverdyan and Nieuwenhuizen that utilizes more rigorous physics for same purpose of extraction of work from a single thermal bath in the quantum regime with quantum coherence. These earlier, original discoverers perhaps have more controversial statements in the article regarding ***free energy extraction***. Using the quantum Langevin equation for quantum Brownian motion, they note that it has a Gibbs distribution only in the limit of weak damping, thus preventing the applicability of equilibrium thermodynamics. The reason is related to quantum entanglement and the necessary mixed state:

> Our main results are rather dramatic, apparently contradicting the second law: We show that the Clausius inequality $dQ \leq TdS$ can be violated, and that it is even possible to extract work from the bath by cyclic variations of a parameter ("perpetuum mobile"). The physical cause for this appalling behavior will be traced back to quantum coherence in the presence of the near-equilibrium bath.[257]

Regarding the ZPF, it is fascinating that *the quantum Langevin equation is a consequence of the fluctuation-dissipation theorem*. (Remember how important I said that equation is?) The authors note that part of the equation includes the fluctuating quantum noise, which has a maximum correlation time and therefore has a long memory (quantum coherence) at low temperature. The Brownian particle of interest also has a semi-classical behavior due to its interaction with the bath, where notably, "there is a transfer of heat, even for T = 0."[258]

The Actual Casimir Effect Equation

In the previous chapters we reviewed the principles of

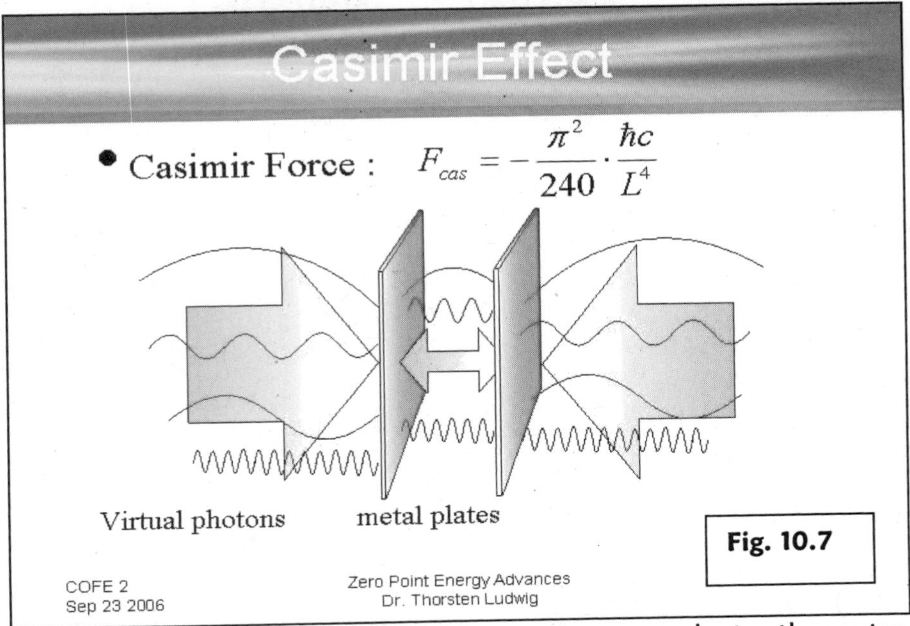

Fig. 10.7

the Casimir force. A passing reference was made to the extra frequencies that are present outside of the cavity and the limited electromagnetic frequencies that are present inside the cavity, between the plates. It is important to realize that just as the Chapter 3 section on Dr. Frank Mead's invention treats ZPE as electromagnetic waves, so can the Casimir experiment be treated.

In Fig. 10.7 we see a slide from Dr. Thorsten Ludwig's presentation at the Second International Conference on Future Energy on "Zero Point Energy Advances" which graphically shows the difference in the electromagnetic frequencies present inside and outside a simple Casimir cavity.

It is only appropriate and fitting that you are also left with the actual simple but elegant equation for the Casimir force (Fig. 10.7). It depends on three constants (Planck's constant, pi, and the speed of light, where $\hbar = h/2\pi$) and of course, the distance L between the plates. There are very few relationships in nature that depend upon the fourth power but here we see one. This causes all of the problems of stiction with such a dramatic increase in the force as the separation distance L gets really tiny. However, the gecko is happy since this effect lets him stick to any surface (even the Science channel recently showed a live gecko in a lab and a close-up of the tiny hairs on its feet which they related to van der Waals forces).

Order Out of Chaos

Mankind has spent a century nurturing the dead remains of dinosaur fossils for fuel and also reaping the archetypal baggage associated with such morbid necrophilia. Now that we are realizing the consequences of our actions and how seriously it will impact the planetary inhabitants for another century to come, it is time to look to a more vital energy source. As an alternative, it can be argued that ZPE is a living energy in the broad sense of the word. The ZPF breathes life into every atom, sustaining its size and shape and it certainly is renewable. Further insights can be gained from the chaotic activity that is at the core of the ZPF. Studying chaos theory, one learns about "strange attractors," "islands of stability" and "order out of chaos." In fact, the Nobel Prize winner, Ilya Prigogine wrote a book with the title, *Order Out of Chaos*, which is an excellent discourse on the topic of this section.[259] When faced with non-linear chaotic systems, like the ZPF, Prigogine makes the case that such systems will often find stability "far from equilibrium." With such a counter-intuitive notion, supported by many specific examples in nature, Prigogine explains the title of his book by saying,

> At all levels, be it the level of macroscopic physics, the level of fluctuations, or the microscopic level, nonequilibrium is the source of order. Nonequilibrium brings 'order out of chaos.'[260]

He also delves into an excellent discussion of entropy and the second law of thermodynamics. Prigogine also explains irreversibility with a re-discovery of the fluctuation-dissipation theorem, without apparently knowing about Callen's earlier publication. "Intrinsic irreversibility is the strongest property: it implies randomness and instability." [261] Thus we have another fundamental description of ZPE from chaos theory, which also includes a description of the amplification of fluctuations as well.[262]

Conclusion

A revolution is taking place in physics spurred by the advances in nanotechnology. As ZPE conversion comes online, which it will sooner than we think, our 20th century "consumption mentality" will be replaced by 21st century "conversion" freedom. Hopefully, the information presented in this book is provocative to the reader, who should check with the references wherever questions might arise. The important part, also true of the whole book, is that ZPE physics is radically different, even in the realm where thermodynamics used to reign supreme. Extracting energy from ZPE has also just become easier, in so many ways, that our future will be much brighter and offer much more freedom for everyone. ZPE energy converters are the true answer to global warming, the oil crisis, third world development, survivalists, space travel, as well as moon and Mars colonies. First, they will start out small, serving the computer battery and cell phone industry. Then, with large scale integration, cars, homes and even industry will be next. Lastly, space travel and space colonies will have a much better voyage in every respect.

Fig. 10.8
Lockheed has shown new interest in funding ZPE research recently perhaps from their commitment to "continuing the voyage"

Questions and Answers

Here are some typical questions that everyone has about zero point energy, especially after reading this book. I have added my answers immediately after each one to help clarify these stumbling blocks.

Q. On page 13 you mention zero point energy as the _kinetic_ energy retained by the _molecules_ of a substance at absolute zero. Two things puzzle me here. One is that if ZPE exists in a vacuum, then presumably no molecules are necessary for it to exist. In a (perfect) vacuum there would be no molecules in which the energy would reside, yet I get the clear impression ZPE would still exist. Also this mention of kinetic energy of molecules gives me as a layperson **the impression that ZPE is *thermal***, a kind of ineradicable residuum of thermal energy of atoms and molecules even at absolute zero. But elsewhere in the book ZPE is defined as *non-thermal*, and being a property of empty space. This seeming contradiction (molecules or no molecules, empty space? thermal or non-thermal?) haunts me throughout the book as "empty" space (*which is clearly not empty*) and absolute zero are mentioned again and again. Wouldn't a perfect vacuum (I know that's theoretical, not real) have no temperature because it would contain no molecules to be in motion (or not, in the case of absolute zero) and thus to create temperature by bouncing around and against each other? (This is probably my ignorance of physics talking, but my ignorance may be typical of the layperson) Maybe a less analytical layman than I won't stumble over it, but I do think it needs to be cleared up, if possible, early on for people who are sensitive to apparent contradictions. I think this seeming contradiction or vacillation was part of what made the book challenging for me to understand. The concept of ZPE itself wouldn't stay put. (Like the fluctuations of the vacuum.)

A. This is a common complaint for anyone grappling with the most contradictory aspects of vacuum physics. The first step toward visualizing what the heck this book is talking about (in terms of ZPE) is to start with the emptying of a container. This is paraphrasing the *Scientific American* article that I refer to in Chapter 2 which helps to step the reader through the stages of understanding that a necessary to grasp the "whole ball of wax." Once the molecules are removed, there is *electromagnetic radiation* to deal with. That's where a clear picture of the traveling wave image repeated twice in the book helps a lot. (This is what engineering and physics students have to memorize in order to pass their courses.) Imagine the EM radiation bouncing all over the inside of the container, at the speed of light, exchanging energy with the sides, as it equalizes the temperature. This is exactly what has been occurring for billions of years across the larger parts of the universe, creating the almost uniform "background radiation" from the

Big Bang or this initial beginnings of the universe as we know it. Therefore, it is not just molecules that create temperature, as astronauts know very well. It is the exchange of EM radiation, mostly in the infrared range that transfers sufficient amounts of energy to cool a substance. Even the human body loses more energy though this type of radiation loss than through and conduction or convection caused by molecules moving. Often in physics, we can only appreciate the experimental or epistemological phenomena first before the ontological noumena presented by the theory can make any sense. This is a fancy way of saying that experimentally, we see Helium remaining a liquid even within 1 degree of absolute zero. At first, the physicist must have exclaimed, "That's impossible!" Then, calculating the amount of ZPE available to the atom, based on the very small size of Helium which can resonate with correspondingly very high frequencies (and therefore, more energy), he probably then appreciated *some type of radiation exchange that is going on*. In other words, "virtual particles" that pop out of the vacuum plenum create (1) heat, (2) pressure, (3) energy, and (4) force which relate to each other. That is a good way to visualize the vacuum dynamics, with some of the pictures provided in this book. Matter gets created out of nothing, once a hole in the continuum is created, bridging the two sides of reality (positive and negative energy realms). All of this can be called "non-thermal" because it really exists even when normal temperature measurements tell us that we have reached the absolute zero-point (-273°C or –459°F).

Q. One question that puzzled me but is of less overall consequence to understanding the book is on one hand the frequent mention of quantum fluctuations being polarized near the boundaries of **charged particles** (e.g., p. 22, 7th line up from bottom) but then at the bottom of page 28 there is a mention of "a polarized vacuum of virtual particles" surrounding protons, **neutrons**, and even micron-sized bacteria. Since neutrons don't have a charge, by definition, it would seem that either the theoretical explanation of the charge as the cause of the Casimir force (polarization as a result of the charge) is erroneous or neutrons should be an exception and should not be surrounded by a polarization of the force.

A. Thanks for pointing out that oversight. The summary in that page region has been corrected to clarify the situation with a **polarized vacuum**, where the oppositely charged virtual particles will tend to be drawn closer to the electric field and the similarly charged virtual particles will be repelled away from the surface (e.g. electrons or protons). When and if the resulting Casimir force is ever measured for such a situation, which is difficult to do with the "Coulomb" interaction compounding the problem, it probably will be about the same as the **non-polarized vacuum** Casimir force where ALL of the virtual particles participate in creating the force, for an electrically neutral surface boundary like a neutron.

Q. What is a **"Casimir cavity"** and what is **the "ground state of hydrogen"?** Is the latter the state in which an electron is in the lowest of several possible orbits, or energy levels, relative to the nucleus (proton)? The Casimir cavity **does** something—it restricts the range of frequencies of ZPE around hydrogen atoms. Does this "range of frequencies" mean the possible orbits of electrons? Or something else? And, what **is** a Casimir cavity (in contrast to what it does) and how does it do it? Is it a literal box at the nanometer level or an energy field or what? I can't conjure an image of it.

A. In regards to this book's theme this is a very important combination question to resolve for the reader. We can look at very small **boxes**, which literally have six walls, or **maybe just two surfaces** as a minimal box with no sides but just a top and bottom, that we can only call a "cavity." Either type of box may be referred to as a cavity in vacuum physics. Now a "Casimir cavity" would be one of a particular size that is less than one micron in width. It could even be spherical or rectangular, as the work by Jordan Maclay has shown. That is the special dimensional region where the Casimir force becomes noticeable. The second part of the question deals with the hydrogen atom and its ground state. This is the lowest energy level that the electron in the hydrogen atom can go to, without touching the nucleus. In quantum mechanics, it is called the "S level" and looks like a fuzzy sphere and is the only level that comes close to what the "Bohr model" tried to propose is the planetary style of "orbiting" that may be part of the electron path, though it behaves as a wave as well. Now if we put the both of these together: the hydrogen atom in the Casimir cavity, there is a quantum vacuum prediction that the real reason that the S level stays away from the nucleus is because of the range of ZPE virtual particles (and their corresponding waves) which are manifesting BETWEEN the nucleus and the electron. Very similar to the work that Koltick is credited for doing with the single electron, the ZPE "cloud" is a full range of waves of virtual particles, ONLY if the particle, or in this case the hydrogen atom, is NOT in a Casimir cavity. As this book shows in many places, a Casimir cavity limits or restricts the wave frequencies that are allowed to exist or manifest between the plates (see Fig. 10.7 for example). This restriction not only **seems to be the primary cause of the Casimir force** but also may affect the ability for the vacuum to push the electron away from the nucleus as well. The "range of frequencies" in the question should not be read into the behavior of the orbits of the electron. This is a summary of the ZPE background frequencies available to the virtual particles as they come into existence momentarily. The range of frequencies are lowered in a Casimir cavity, while outside of it, the space has the full crushing weight of all of the frequency guns hammering away. I am reminded of a **science experiment** that anyone can do at home which is **similar to the Casimir cavity** experiment: Fill a glass to overflowing with water. Slide a piece of cardboard, paper or stiff lightweight plastic, slightly bigger than the glass, over the top of it, leaving absolutely no air bubbles. With two hands, carefully turn the glass upside down. Slowly release the hand holding the cardboard and hold it up by one hand only on the glass. The force of gravity pulling the water down is overcome by the vacuum force keeping the water in the glass. The other explanation often used is that the air pressure on the cardboard is pushing the water upwards with fourteen pounds per square inch, which is much more than the small amount of water in the glass weighs, per square inch. This is very similar to the Casimir cavity phenomenon where forces on the outside are pushing against the surface with more force than is available inside.

Q. Does the fact about one's **uncertainty** about momentum illuminate the nature of zero point energy? That's not clear to me. The first longish quote on page 36 attempts to answer this by saying at the end that "the energy associated with the uncertainty in momentum gives the zero point energy." Huh? Why does extreme uncertainty about momentum "give" energy? Doesn't it just allow the possibility of energy? Since I have almost infinite uncertainty about all this stuff my energy level must be approaching infinity.

A. The simple answer to this question about uncertainty forces some of the most important concepts underlying quantum mechanics to come to the surface. Whether we look at the pairs (for physics reasons) of energy and time, or position and momentum, these pairs of "opposites" behave as it they are tugging at each other. Keeping the measure of the product of their uncertainty above a small number that depends on Planck's constant is a rule that nature has imposed upon us. What quantum physicists like Peter Milonni and others have deduced from the all-pervading influence that such a principle or rule has on everything is that it points to the underlying TURBULENCE that causes the uncertainty in the first place. Moreover, there is precise order in the chaos, so much so that we can literally DERIVE the primary energy level of ZPE that is available throughout the universe directly from the uncertainty principle, as I have done for the reader on page 205. It is not so much whether we pick the momentum and position pair or the energy and time pair (there is a third choice too). The important part is the $h/4\pi$ which is nature's limit, analogous to the speed of light c which is also regarded as another one of nature's limits. The surprise, which is proven on page 205, is that the chaotic uncertainty itself has as its basis the very ZPE that we are studying in this book. The two are literally the same phenomenon. The uncertainty principle is really another way of stating the effects of ZPE in every day laboratory physics experiments.

Appendix

Exciting Papers on Zero-Point Energy

(The easy-to-read ones are up front)

Glossary

ZPE & Casimir Web Links

Index

References

Book Reviews

Report on Tom Valone's Lecture on Zero-Point Energy Extraction from the Quantum Vacuum

Review by T. Cullen, Sterling D. Allan, Susan Carter *Pure Energy Systems*, August, 2004
http://www.pureenergysystems.com/events/conferences/2004/teslatech_SLC/TomValone/ZPE_Extraction_QuantumVacuum.htm

In a talk titled "Feasibility of Zero-point Energy Extraction from the Quantum Vacuum for Useful Work", Dr. Valone reviewed what ZPE is and how it is being used as he addressed the ExtraOrdinary Technology conference in Salt Lake City on Friday, July 30, 2004, .

To kick off his talk, he cited a recent article in the *Los Angeles Times* (July 25, 2004) that stated that our only hope for solving the energy crisis is to be willing to consider "extreme possibilities." With that introduction, he then cited numerous instances in which mainstream journals of science have been dabbling in a discussion of zero point energy.

He discussed how the vacuum used to be considered empty, but now it has been shown that the vacuum contains an enormous amount of energy. Even when you remove all sources of energy and cool a region to very close to absolute zero (the zero point), there is useable energy present in abundance. He said that this is why Helium stays liquid at fractions of a degree Kelvin. Dr. Valone described an experiment by Koltick that shows the effect of virtual particle "dressing" that shrouds an electron. The *Quantum Vacuum* text by Milonni, he says, estimates the ZPE energy density at 220 erg/cc in optical regions. These measurements were able to be made because of science's ability to study matter at the nanoscopic level. He also reported that gravity and inertia are proven to be effects of ZPE, by none other than Dr. Hal Puthoff at the Institute for Advanced Studies at Austin.

Dr. Valone described the Casimir Effect, and how it can be used to tap ZPE. This is the slight attraction seen in metal plates when placed very close (atomic distances) to each other. The attraction can be shown to come from ZPE. Valone cites evidence that the ZPE is not conserved, and does not follow the normal laws of energy conservation. He also showed documented research that sometimes the Casimir Force is repulsive due to magnetic or thermal conditions. In some cases, the force changes sign as the temperature increased. This can lead to ways to manipulate and control ZPE.

In his presentation, Dr. Valone cited several mainstream science journals that are now publishing works by ZPE researchers (Some examples: *Phys. Review Letters* #92, 2004; *Aviation Week* March 1 2004; *Science* v. 299 issue 5608: 2003 p. 862.) He talked about Robert L. Forward's early work proposing to extract energy from the Casimir Effect and how he made an electron storage battery instead. He said that now that we have the ability to work with nanotechnology, we have the tools to extract energy from the zero point field.

Valone also discussed toroidal EM fields. He said that a ZPE field loses its drag when the temperature nears 0ºK, according to Froning's research. He also cited references that show that we can now explain how it is possible to extract useable energy from a single heat source (not from a temperature difference), which challenges the First Law of Thermodynamics.

Valone also talked about quantum coherence and micro laser cavities. He cited the works of Pinto (*Phys Rev. B* 60, 21, 1999 p. 4457) and how he was able to use a micro laser cavity to change the properties and increase the Casimir force – like turning on a light and getting a force out. He also spoke about the fluctuation driven electricity experiments by Crooks (*Phys Rev. E.* 60, 1999) where he is able to get motion from zero input force. He described this as like a "quantum ratchet". The research by Linke in *Science* magazine was also cited. Valone described a report in the July 8, 2004 issue of *Nature* about how the "Dark Energy" of astronomy is ZPE, and is why the universe is accelerating. He said that when he confronted astronomer Reba Goodman about this nomenclature, and that they are describing Zero Point Energy, so why didn't they just call it ZPE, the astronomer replied that they "wanted to keep it vague."

He said spectral density for ZPE is Plank's 2nd radiation law, which has now been also measured in superconducting tunnel diode noise and reported in a journal article entitled "Has Dark Energy Been Measured in the Laboratory?" Where superconducting circuits are concerned, it is interesting to point out that "Perpetual Motion Machines of the Third Kind," as he put it, have been achieved with superconducting currents that won't stop, even after ten years of operating with no further energy input, such as those used in MRI machines.

Valone listed the following patents as the most significant in ZPE research: "Rectifying Thermal Electric Noise" by Charles Brown 3890161, and 4004210 by Yater; and 4704622 by Capasso, which actually acknowledges ZPE. He mentioned that metal-metal nanodiodes probably hold the key to ZPE usage with millipore sheets.

He also cited the work of Yasamoto, et. Al. (2004, *Science*, 304:1944) covering peptide molecular photodiodes just 1 nm across -- another example of a molecular tool for studying this zero point energy that shows up on the molecular level.

Dr. Valone's report makes one realize that experiments tapping ZPE are now starting to be researched and discussed by the respected scientific community's peer-reviewed journals.

Website:

- www.IntegrityResearchInstitute.org of which Dr. Valone is President.

Related Works by Tom Valone:

- Feasibility Study of Zero-Point Energy Extraction from the Quantum Vacuum for the Performance of Useful Work (3 Mb doc) [HTML] 2004
- Inside Zero Point Energy - 1999 – www.seaspower.com/InsideZeroPoint-Valone.htm
- Zero Point Energy and the Future - 2003 – http://erols.com/iri/orderform14.html
- Zero Point Energy: The Fuel of the Future - forthcoming book

Referenced Patent of note

- U.S. 3,890,161, a chip which absorbs heat directly while producing electrical power

Event Sponsors

- TeslaTech - www.teslatech.info
- Institute of New Energy - www.padrak.com/ine
-

Why Declassifying and Releasing Zero Point Energy Technology Will Not Cause The Sky To Fall

by Richard Boylan, Ph.D.

Before his untimely death in 2000, Dr. Michael Wolf of the Special Studies Group (SSG) of the U.S. National Security Council engaged in dialogues with me about the challenges inherent in formal governmental acknowledgment of the reality of UFOs and Star Visitor contact. One of the chief impediments to the SSG's releasing such information was the fear by some Group members that such a release would promptly lead to access to knowledge about Zero Point Energy technology, which the Star Visitors have long utilized.

Zero Point Energy refers to an omnipresent, invisible-but-tappable field of energy in the electromagnetic quantum field. Properly tapped, such energy offers the potential of almost-free, virtually-unlimited nonpolluting energy to power the motors and devices which make up our technologically-advanced civilization.

Dr. Wolf said that some SSG members feared that the public release of ZPE technology would lead to a collapse of the global economy, petroleum-based as it largely is. I argued with him that those SSG members were wrong. And here is why. There will be no sudden loss-of-demand for petroleum products after Star Visitor-based Zero Point Energy technology is declassified and made public.

The history of introduction of new technologies is more gradualistic than in fear-based scenarios. People are not going to junk a car they have sunk $25K into because ZPE is announced. Oil stations will not close suddenly, with all those existing gas-burners still on the road. It will take years to design and mass-produce practical, individual-scale ZPE units. It will take years to gradually convert production lines from petroleum-based energy producers (coal-burning power stations, internal-combustion generators, etc.) and energy-consumers (cars, boats, airplanes, etc.) to ZPE-based equivalents. As with any shift, the rich will get the first units. As production widens, and the price of ZPE-based products falls, the masses will get at the new ZPE-based products as their old petro-based ones wear out. Workers will continue producing petro-based products until the demand gradually declines to nothing. Other workers will get the new jobs in new ZPE industries, and those existing industries modernizing into ZPE, springing up to produce ZPE products, as the demand for new ZPE products ramps up.

As for the petrodollar-based sheikdoms and other countries that have relied on oil reserves as principal cash flow, they will have several strategies to follow. Many will take their continuing (but gradually-declining) petrodollar earnings and invest in the newfangled ZPE industries. As those ZPE industries take off, those former-oil investor countries will now realize profits and increasing revenues from owning major shares in ZPE industries, or in hosting some of these ZPE industries. Other former-petro countries will use their continuing but declining petrodollar earnings to revamp their economies to emphasize and expand other trade and production items that remain viable income sources, such as cocoa, hemp, emeralds, fruit, construction materials, industrial chemicals, and indigenous crafts. Still others will revamp to start up new industries, like hosting world tourism (much more possible in a ZPE world), or authentic shamanic mental-development and health seminars and mentoring internships utilizing some of their previously-ignored indigenous populace.

Yes, with abundance, as people no longer have to toil all day to feed, clothe and house themselves, they will have more time to turn to cultivation of their inner, metaphysical and spiritual life. And as personal economics levels globally, the average income will be roughly what is now upper-working-class, meaning everyone will have some disposable income to spend beyond the necessities. This is a huge increase in market demand for goods and services. And such goods and services will be available from a much-wider array of locations, as ZPE goes worldwide.

What is missing out of most analyses which superimpose the introduction of Zero Point Energy technology onto the existing industrialized world is the empowering valence which widely-available ZPE has for narrowing the standard-of-living gap between developed and have-not countries.

ZPE availability will make many underdeveloped countries, which currently rely on imports and loans to sustain their population, to become self-sufficient producers. With a ZPE unit the size of a packing crate able to produce unlimited low-cost water-pumping and electricity for an entire village in every Third World country, those villagers will be in a position to produce their own food and fibre hydroponically (even in desert or arctic climates), and utilize Appropriate-Technology-sized machines to manufacture housing material, communication devices, and the other tools of modern society.

And as countries everywhere succeed in feeding, clothing, housing, educating, and engaging all their citizens in productive service, the threat of territorial wars for raw materials and more land with its resources will cease to have any reason for existence. At that point it will then also be safe to release to the public full information on antigravity technology, which the SSG has currently sequestered. (Some within the SSG fear that the superweapons aspects of antigravity technology need to be embargoed until peace prevails in the world. ZPE technology release will hasten that day.) Then, widespreadly available antigravity technology will further enhance the global exchange of goods and services in a ZPE-empowered world.

We do not need planet-killing petroleum-based energy to sustain our world. We DO need the immediate declassification and release of Zero Point Energy technology, so that we can start ramping up our present world to a truly-civilized global society. Ecologically sound. Socially just. Economically sustainable in the long term. Peaceful as only economic sufficiency can ensure peace. And finally fit to walk out and accept our role as the latest world accepted into the federation of cosmic cultures as a truly civilized planet.

Richard Boylan, Ph.D., LLC, P.O. Box 22310, Sacramento, CA 95822, USA

Phone, voice mail, fax number: (916) 422-7400

E-mail: drboylan@sbcglobal.net Website: www.drboylan.com

Zero-Point Energy Shows Promise for Its Use

Thomas Valone, PhD
Integrity Research Institute, Washington DC
www.IntegrityResearchInstitute.org
Presented to Institute for New Energy, USPA Conference, and Conference on New Energy Alternatives

Introduction

A recently published zero-point energy (ZPE) study,[1] marks a new dimension in research directions for our fuelless energy future, which this author believes is essential for the survival and travel independence of the human race. A number of significant discoveries were made with the study, by interpreting little-known journal articles with an engineering focus toward energy and propulsion applications.

Zero-Point Energy Primer

Zero-point energy (ZPE) is a universal natural phenomenon of great significance which has evolved from the historical development of ideas about the vacuum. In the 17th century, it was thought that a totally empty volume of space could be created by simply removing all gases. This was the first generally accepted concept of the vacuum. Late in the 19th century, however, it became apparent that the evacuated region still contained thermal radiation. To the natural philosophers of the day, it seemed that all of the radiation might be eliminated by cooling. Thus evolved the second concept of achieving a real vacuum: cool it down to zero temperature after evacuation. Absolute zero temperature (-273C) was far removed from the technical possibilities of that century, so it seemed as if the problem was solved. In the 20th century, both theory and experiment have shown that there is a non-thermal radiation in the vacuum that persists even if the temperature could be lowered to absolute zero. This classical concept alone explains the name of "zero-point" radiation.

---

[1] Valone, Thomas, *Practical Conversion of Zero-Point Energy: Feasibility Study of the Extraction of Zero-Point Energy from the Quantum Vacuum for the Performance of Useful Work*, Integrity Research Institute, 2005

In 1891, the world's greatest electrical futurist, Nikola Tesla, stated,

> Throughout space there is energy. Is this energy static or kinetic? If static our hopes are in vain; if kinetic – and we know it is, for certain – then it is a mere question of time when men will succeed in attaching their machinery to the very wheelwork of Nature. Many generations may pass, but in time our machinery will be driven by a power obtainable at any point in the Universe.

The subject of zero-point energy is presently being tackled with appreciable enthusiasm and it appears that there is little disagreement that the vacuum could ultimately be harnessed as an energy source. Indeed, the ability of science to provide ever more complex and subtle methods of harnessing unseen energies has a formidable reputation.

A good experiment proving the existence of ZPE is accomplished by cooling helium to within microdegrees of absolute zero temperature. It will still remain a liquid. Only ZPE can account for the source of energy that is preventing helium from freezing.

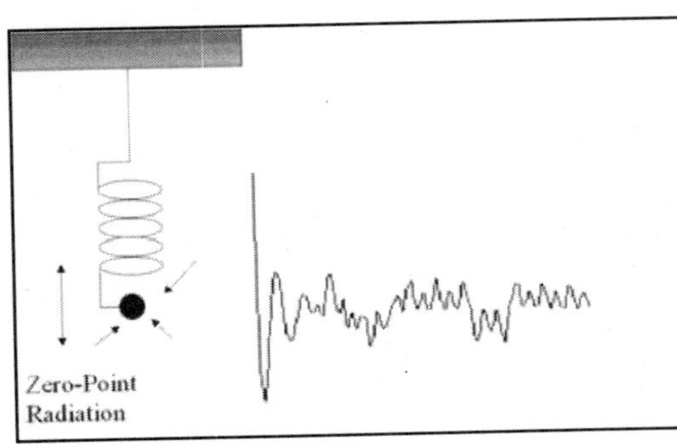

Besides the classical explanation of zero-point energy referred to above, there are rigorous derivations from quantum physics that prove its existence. It is possible to get a fair estimate of the zero point energy using the uncertainty principle alone. Planck's constant h (6.63×10^{-34} joule-sec) offers physicists the fundamental size of the quantum. It is also the primary ingredient for the uncertainty principle, often seen as the minimum uncertainty of position x and momentum p: $\Delta x \, \Delta p \geq h/4\pi$.

In quantum mechanics, Planck's constant also is present in the description of particle motion. The harmonic oscillator reveals the effects of zero-point radiation on matter. The oscillator consists of an electron attached to an ideal, frictionless spring. When the electron is set in motion, it oscillates about its point of equilibrium, emitting electromagnetic radiation at the frequency of oscillation. The radiation dissipates energy, and so in the absence of zero-point radiation and at a temperature of absolute zero the electron eventually comes to rest. Actually, zero-point radiation continually imparts random impulses to the electron, so that it never comes to a complete stop [as seen in Figure]. Zero-point radiation gives the oscillator an average energy equal to the frequency of oscillation multiplied by one-half of Planck's constant.

Summary of New ZPE Findings

Based on the level of agreement between theory and experiment in ZPE conversion, exploitation of zero-point energy extraction should begin with the nanoscopic realm. In particular, it is recommended that research be initiated, encouraged and supported in the following promising areas:

1) metal-metal diodes should be researched, with attention to the Johnson noise voltage and purported lack of diode barrier, along with the possible mass production of high density substrates;
2) ratchet and ratchetlike asymmetries should be researched, by government, industry and academia, so that a TB lattice or diode assembly may one day offer a truly solid state transducer for ZPE;
3) quantum coherence, refractive index change, and stochastic resonance research should continue with a goal of reducing the present relatively large energy investment, so that more robust avenues of product development in ZPE thermodynamics may be achieved. Brownian motors, thermal fluctuation rectifiers, and quantum Brownian nonthermal rectifiers utilizing AQSR have already achieved a level of theoretical and experimental confidence where further physics research and engineering studies can offer fruitful rewards in the production of rectified DC electricity. This mode of ZPE conversion research and development needs to be continued with earnest in order

to expand mankind's woefully limited portfolio of energy choices.

A broad outline of how to undertake the recommended development work necessarily has specific tasks and milestones associated with

a) the confirmation of ZPE quantum effects described in this study on a larger scale;
b) replication of results but also optimization of results; and
c) engineering tasks of conductor and semiconductor design, nanowires and ohmic contacts.

All of these, along with other tasks not mentioned, need to be included. The project would also include estimates of output current and energy production with any given geometry. Parallel development paths in research and development will always accelerate the completion of the optimum design. A market study should also accompany the work, so a clear focus on the existing niche to be filled is maintained. A national or international project proposal that estimates the required project scope, resources, break-even point and identifies major milestones, has to be formulated, if major progress in ZPE usage is to be achieved. Simply commissioning another study to follow up this study will lead only to institutionalizing the effort without accomplishment of set goals.

This feasibility study of ZPE extraction for useful work has presented a balanced and detailed assessment with scientific integrity, engineering utility and the likelihood of success for further development. It is justifiably concluded that *zero-point energy is deserving of more attention by engineers and entrepreneurs as a serious and practical energy source for the near future*. The proposed project plan for ZPE development, yet to be written, has been reduced to a business endeavor and an exercise in return on investment. A few amazing examples of some recent ZPE discoveries follow below.

Focusing Vacuum Fluctuations

A method has now been found to concentrate and focus ZPE.

In a Casimir Workshop held in 2002 at Harvard University, L.H. Ford and N.F. Svaiter (Physics Department, Tufts University, Medford, MA 02155) found that the quantization of the electromagnetic field in the presence of a parabolic mirror is possible in the context of a geometric optics approximation. They calculate the mean squared electric field near the focal line of a parabolic cylindrical mirror. This quantity is found to grow as an inverse power of the distance from the focus. Ford and Svaiter give a combination of analytic and numerical results for the mean squared field. In particular, they find that the mean squared ***electric field can be either negative or positive***, depending upon the choice of parameters. The case of a negative mean squared electric field corresponds to a repulsive Van der Waals force on an atom near the focus, and to a region of negative energy density. Similarly, a positive value corresponds to an attractive force and *a possibility of atom trapping in the vicinity of the focus.*

Movement from Nothing

Empty space can set objects in motion, a physicist claims.[2]

Motion can be conjured out of thin air, according to a physicist in Israel. Alexander Feigel of the Weizmann Institute of Science in Rehovot says that objects can achieve speeds of several centimetres an hour by *getting a push from the empty space of a vacuum*[3].

---

[2] Philip Ball, *Nature*, Feb. 2004
http://www.nature.com/Physics/Physics.taf?g=&file=/physics/highlights/6974-3.html&filetype=&_UserReference=C0A804F54651F06AE1CBD407899240295C0F

[3] Feigel, A. Quantum vacuum contribution to the momentum of dielectric media. *Physical Review Letters*, **92,** 020404, doi:10.1103/PhysRevLett.92.020404 (2004)

No one has yet measured anything being set in motion by emptiness. But Feigel thinks it should theoretically be possible to make use of the effect to shunt tiny amounts of liquids around on a lab chip, for example. Such small-scale experiments could be useful for chemists interested in testing thousands of different drugs at the same time, or for forensic scientists who need to do analyses on tiny amounts of material.

Empty space contains energy from virtual particles which can move objects.

The whole idea of getting movement from nothing sounds like a gift to advocates of perpetual-motion machines. But there's nothing in Feigel's theory that violates the fundamental laws of physics, so this doesn't provide a way to cheat the Universe and get free energy.

Instead, Feigel draws on the well-established notion that empty space does contain a little bit of energy. This 'vacuum energy' is a consequence of the uncertainty principle — one of the cornerstones of quantum mechanics.

Because of the uncertainty principle, subatomic particles or photons can appear spontaneously in empty space — provided that they promptly vanish again. This constant production and destruction of 'virtual particles' in a vacuum imbues the vacuum with a small amount of energy.

Moving in a Vacuum

Feigel considered the effects of virtual photons on the momentum — a property defined as mass multiplied by velocity — of objects placed in a vacuum, and came to a surprising conclusion. He started with the fact that electrical and magnetic forces between objects are mediated by photons that flit between them. So an object placed in strong electric and magnetic fields can be considered to be immersed in a sea of these transitory, virtual photons.

Feigel then showed that the momentum of the virtual photons that pop up inside a vacuum can depend upon the direction in which they are travelling. He concludes that if the electric field points up and the magnetic field points north, for example, then east-heading photons will have a different momentum from west-heading photons. So the vacuum acquires a net momentum in one direction — it's as though the empty space is 'moving' in that direction, even though it is empty.

It is a general principle of physics that momentum is 'conserved' — if something moves one way, another thing must move the other way, as a gun recoils when it shoots a bullet. So when the vacuum acquires some momentum from these virtual photons, the object placed within it itself starts to move in the opposite direction.

Feigel estimates that in an electric field of 100,000 volts per metre and a magnetic field of 17 tesla — both big values, but attainable with current technology — an object as dense as water would move at around 18 centimetres per hour.

General Relativity and Vacuum Energy

Jordan Maclay, a NASA-funded quantum researcher, has some interesting observations about ZPE that show its power and complexity.[4] In general relativity, he notes, any form of energy has an equivalent mass, given by $E = mc^2$, and is therefore coupled to gravity. This enormous zero-point energy density is equivalent to a mass density of about 10^{92} kg/cc, and would be expected to cause an enormous gravitational field. This large field leads to some major problems with general relativity, such as the collapse of the universe into a region of space that is about 1 Planck length across. Thus we have an inconsistency in two very important and well-verified theories, QED and General Relativity. A brief discussion of this problem is given in the excellent book *Lorentzian Wormholes* (Springer-Verlag, 1996, p. 82) by Matt Visser.

The ZPE in a region of space the size of a proton is equivalent to the mass of the universe.

As an instructional exercise, it is possible to compute the equivalent mass for a region of the vacuum about the size of a proton, which is approximately a sphere about 10^{-13} cm across, using the enormous energy density formally predicted above. This process yields an equivalent mass of about 10^{53} kg. This means the vacuum energy contained within a region of

[4] Jordan Maclay, Quantum Fields, LLC www.quantumfields.com

space the size of a proton is equivalent to a mass of about 10^{53} kg. A very rough estimate of the number of nucleons in the universe is 10^{80}. This number is based on the statistical distribution of stars in galaxies and the number of galaxies. Most of the mass of matter is in nucleons, so the mass of the universe is roughly the weight of a proton times 10^{80} or about 10^{53} kg, which is the same as the mass equivalent of the vacuum energy in a region the size of a proton. Conclusion: A volume the size of a proton in empty space contains about the same amount of vacuum energy as all the matter in the entire universe! This sounds like the poet's words, which now ring true, "To see the world in a grain of sand."

Brownian Motor

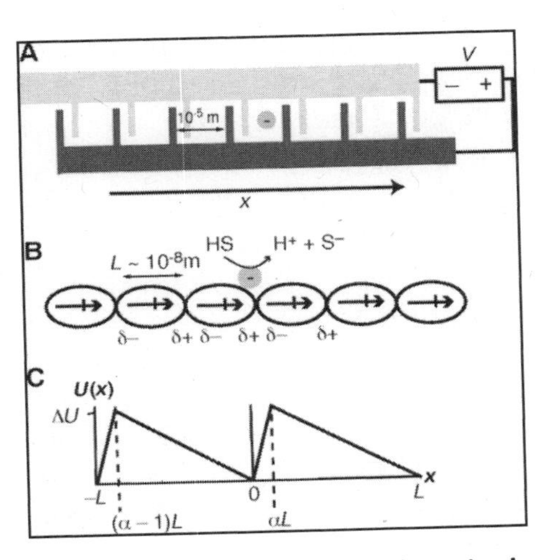

R. Dean Astumian has proposed experiments for nonequilibrium fluctuations, whether generated externally or by a chemical reaction far from equilibrium, that can bias the Brownian motion of a particle in an anisotropic medium without thermal gradients, a net force such as gravity, or a macroscopic electric field.[5] Fluctuation-driven transport is one mechanism by which chemical energy can directly drive the motion of particles and macromolecules and may find application in a wide variety of fields, including particle separation and the design of molecular motors and pumps. Recent work has focused,

> With no net input force, a tiny Brownian motor can transport particles, thus performing work.

[5] Astumian, R.D., "Thermodynamics and Kinetics of a Brownian Motor," *Science*, 276, 1997, p. 5314

however, on the possibility of an energy source other than a thermal gradient to power a microscopic motor. If energy is supplied by external fluctuations or a nonequilibrium chemical reaction, Brownian motion can be biased if the medium is anisotropic, even in an isothermal system. Thus, directed motion is possible without gravitational force, macroscopic electric fields, or long-range spatial gradients of chemicals.

In devices based on biased Brownian motion, net transport occurs by a combination of diffusion and deterministic motion induced by externally applied time-dependent electric fields. Although the electric generator is certainly a macroscopic device, *the electric field in the x direction averaged over a spatial period is zero no matter what the voltage, and so there is no net macroscopic force.*[6] A nonequilibrium fluctuation can be produced by using a switching device that imposes an externally defined but possibly random modulation of the voltage. Recent experiments have shown that unidirectional motion of microscopic particles can be induced by modulating the amplitude of such an *anisotropic sawtooth potential*. Theory shows that the direction of flow is governed by a combination of the local spatial anisotropy of the applied potential, the diffusion coefficient of the particle, and the specific details of how the external modulation is carried out.

The recent work on fluctuation-driven transport leads to optimism that similar principles can be used to design microscopic pumps and motors – machines that have typically relied on deterministic mechanisms involving springs, cogs, and levers – from stochastic elements modeled on the principles of chemical reactions and noise-assisted processes.

Energy from a Single Heat Bath

Nothing gets the classical physicist's goat more than asserting that energy can be drawn from a single heat source, in apparent violation of thermodynamics. However, that is precisely what now seems

---

[6] H. Linke, et al., "Experimental Tunneling Ratchets," *Science* – 286 (5448), 1999, p. 2314; also see L. P. Faucheux, et al., *Phys. Rev. Lett.* **74**, 1504 (1995)

perfectly feasible: a quantum Carnot engine in which the atoms in the heat bath are given a small bit of *quantum coherence*. The induced quantum coherence becomes vanishingly small in the high-temperature limit at which they operate and the heat bath is essentially thermal. However, the phase , associated with the atomic coherence, provides a new control parameter that can be varied to increase the temperature of the radiation field and to extract work from a single heat bath. The deep physics behind the second law of thermodynamics is not violated; nevertheless, the quantum Carnot engine has certain features that are not possible in a classical engine.[7] This invention is a stellar example of quantum physics paving new ground in the inevitable conversion of ZPE for macroscopic use.

In the presence of unbiased, asymmetric forcing, a noise-directed current always occurs in a dissipative tight-binding (TB) lattice, because of the ratchet-like effect of the asymmetric forcing, like the stochastic ratchets that rectify thermal noise. With stochastic resonance, nonthermal fluctuations are effectively rectified, creating a measurable current. Goychuk believes that the effect should be already observable in superlattices and/or optical lattices.[8] Anomalous transport properties, using SR, which do not exploit the ratchet mechanism have been investigated in driven periodic tight-binding lattices near zero DC bias with the combined effects of DC and AC fields, or DC field and external noise. In particular, Goychuk et al. have found that periodic TB lattices can be driven by unbiased nonthermal noise generated from the vacuum ZPF, generating an electrical current as a result of a "ratchetlike mechanism," as long as there is quantum dissipation in the system.

For reference, it is worth mentioning that in crystal lattices, thermal fluctuations appear at environmental temperatures, with $\frac{1}{2}m v_o^2 \langle u^2 \rangle = 3(\frac{1}{2} k_B T)$ energy level where m and v_o are the mass and frequency of the harmonic oscillations and u is the displacement from a fixed lattice site. The nonthermal oscillations associated with

---

[7] Marlan O. Scully, "Extracting Work from a Single Heat Bath via Vanishing Quantum Coherence"
Science, Vol. 299, Issue 5608, February 7, 2003, p. 862-864

[8] Goychuk, I. et al., "Nonadiabatic quantum Brownian rectifiers" *Physical Review Letters*, Vol. 81, No. 3, 1998, p. 649

ZPE are $mv_o^2\langle u^2\rangle = 3(\frac{1}{2}\hbar v_o)$ in terms of energy, adding to the lattice thermal fluctuations.

With the introduction to aperiodic quantum stochastic resonance (AQSR) along with the rectification of nonthermal noise, it makes sense to investigate the amplification of quantum noise. Milonni points out that "the vacuum field may be amplified...if the spontaneously emitted radiation inside the cavity is amplified by the gain medium, then so to must the vacuum field entering the cavity. Another way to say this is that 'quantum noise' may be amplified." Since the SR TB lattice current output depends on the noise level, as in the Goychuk simulation, the optimum level of energy extraction depends on parameter control, as in quantum optics, which utilizes quantum noise amplification. This is similar to amplified stimulated emission (ASE) which also uses a gain medium.

Conclusion

A few pioneers have proposed that a solid state diode or diode array can serve as the template for a ZPE converter. For example Joseph Yater, with his patents and *Physical Review* articles, was one of these notable visionaries. He theorized that a Schottky diode, formed between a semiconductor and a metal, with nonlinear rectifying characteristics and fast switching speeds, could be the diode of choice for rectifying thermal noise. Yater notes in his U.S. patent #4,004,210 that "for the long range design goals, sub-micron circuit sizes are required if all the high power goals of megawatts per square meter are to be achieved." To this I would add the limitation of "zero-bias" diodes, so that a maximum of random electron fluctuations can be converted or "rectified" into DC electricity.

In regards to *rectifying thermal electrical noise*, it is worth mentioning the U.S. Patent #3,890,161 by Charles M. Brown that utilizes an array of nanometer-sized metal-metal diodes, capable of rectifying frequencies up to a terahertz (10^{12} Hz). Brown notes that thermal agitation electrical noise (Johnson noise) behaves like an external signal and can be sorted or preferentially conducted in one direction by a diode. The Johnson noise in the diode is also generated at the junction itself and therefore, requires no minimum signal to initiate the conduction in one direction. The thermal noise voltage is normally given by $V^2 = 4k_BTRB$ where R is the device resistance and B is the bandwidth in Hertz. Brown's diodes also require no external power to operate, in contrast to the Yater diode invention. Brown also indicates that heat is absorbed in the system, so that a cooling effect is noticed, because heat (thermal noise) energy energizes the carriers in the first place and some of it is converted into DC electricity. In contrast, the well-known Peltier effect is the closest electrothermal phenomenon similar to this but requires a significant current flow into a junction of dissimilar metals in order to create a cooling effect (or heating). Brown suggests that a million nickel-copper diodes formed in micropore membranes, with sufficient numbers in series and parallel, can generate 10 microwatts. The large scale yield is estimated to be several watts per square meter.

Taking these two inventions as a starting point for hardware, the transition to engineering quantum Brownian *nonthermal rectifiers* can be much smoother for the nanophysicist or nanotechnician. For example, as this summary article is being drafted the latest news in this ongoing development is that a molecular photodiode rectifier has been invented, which meets some of the characteristics required for ZPE conversion.[9]

Resources

The most comprehensive textbook covering the Casimir effect and related vacuum phenomena is *The Quantum Vacuum*, by Peter Milonni, Academic Press(1994). The Casimir effect is covered in *Quantum Mechanics*, Leslie Ballentine, Prentice-Hall (1990). The most recent resource paper with many references on Casimir forces was published by S. Lamoreaux, *American Journal of Physics*, vol. 67, pp. 850-861, October, 1999. A good review describing the breadth of Casimir phenomena is "Casimir Forces" by Peter Milonni and Mei-Li Shis, in *Contemporary Physics*, vol. 33, 313-322 (1992). Another informative review is E.Elizalde and A.Romero, "Essentials of the Casimir Effect and its Computation," *Am. J. Phys.* **59**, 711-719 (1991). A long and detailed review is given by P. Plunian, B. Muller, W. Greiner, "The Casimir Effect," *Physics Reports* (Review Section of *Physics Letters*) **134**, 2&3, pp. 87-193 (1986). Much information is contained in the text *The Casimir Effect and its Applications* by V. Mostepanenko and N. Trunov, published by Oxford University Press, 1997.

---

[9] S Yasutomi et al. 2004 *Science* 304, p. 1944

Zero Point Energy and Zero Point Field

Bernard Haisch,PhD
Calphysics Institute
http://www.calphysics.org/zpe.html

Introduction

Quantum physics predicts the existence of an underlying sea of zero-point energy at every point in the universe. This is different from the cosmic microwave background and is also referred to as the electromagnetic quantum vacuum since it is the lowest state of otherwise empty space. This energy is so enormous that most physicists believe that even though zero-point energy seems to be an inescapable consequence of elementary quantum theory, it cannot be physically real, and so is subtracted away in calculations.

A minority of physicists accept it as real energy which we cannot directly sense since it is the same everywhere, even inside our bodies and measuring devices. From this perspective, the ordinary world of matter and energy is like a foam atop the quantum vacuum sea. It does not matter to a ship how deep the ocean is below it. If the zero-point energy is real, there is the possibility that it can be tapped as a source of power or be harnessed to generate a propulsive force for space travel.

The propeller or the jet engine of an aircraft push air backwards to propel the aircraft forward. A ship or boat propeller does the same thing with water. On Earth there is always air or water available to push against. But a rocket in space has nothing to push against, and so it needs to carry propellant to eject in place of air or water. The fundamental problem is that a deep space rocket would have to start out with all the propellant it will ever need. This quickly results in the need to carry more and more propellant just to propel the propellant. The breakthrough one wishes for deep space travel is to overcome the need to carry propellant at all. How can one generate a propulsive force without carrying and ejecting propellant?

There is a force associated with the electromagnetic quantum vacuum: the Casimir force. This force is an attraction between parallel metallic plates that has now been well measured and can be attributed to a minutely tiny imbalance in the zero-point energy in the cavity between versus the region outside the plates. This is not useful for propulsion since it symmetrically pulls on the plates. However if some asymmetric variation of the Casimir force could be identified one could in effect sail through space as if propelled by a kind of quantum fluctuation wind. This is pure speculation.

The other requirement for space travel is energy. A thought experiment published by physicist Robert Forward in 1984 demonstrated how the Casimir force could in principle be used to extract energy from the quantum vacuum (Phys. Rev. B, 30, 1700, 1984 </articles/Forward1984.pdf>). Theoretical studies in the early 1990s (Phys. Rev. E, 48, 1562, 1993 </articles/CP93.pdf>) verified that this was not contradictory to the laws of thermodynamics (since the zero-point energy is different from a thermal reservoir of heat). Unfortunately the Forward process cannot be cycled to yield a continuous extraction of energy. A Casimir engine would be one whose cylinders could only fire once, after which the engine become useless.

Origin Of Zero-Point Energy

The basis of zero-point energy is the Heisenberg uncertainty principle, one of the fundamental laws of quantum physics. According to this principle, the more precisely one measures the position of a moving particle, such as an electron, the less exact the best possible measurement of momentum (mass times velocity) will be, and vice versa. The least possible uncertainty of position times momentum is specified by Planck's constant, h. A parallel uncertainty exists between measurements involving time and energy. This minimum uncertainty is not due to any correctable flaws in measurement, but rather reflects an intrinsic quantum fuzziness in the very nature of energy and matter.

A useful calculational tool in physics is the ideal harmonic oscillator: a hypothetical mass on a perfect spring moving back and forth. The Heisenberg uncertainty principle dictates that such an ideal harmonic oscillator -- one small enough to be subject to quantum laws -- can never come entirely to rest, since that would be a state of exactly zero energy, which is forbidden. In this case the average minimum energy is one-half h times the frequency, hf/2.

Radio waves, light, X-rays, and gamma rays are all forms of electromagnetic radiation. Classically, electromagnetic radiation can be pictured as waves flowing through space at the speed of light. The waves are not waves of anything substantive, but are in fact ripples in a state of a field. These waves do carry energy, and each wave has a specific direction, frequency and polarization state. This is called a "propagating mode of the electromagnetic field."

Each mode is subject to the Heisenberg uncertainty principle. To understand the meaning of this, the theory of electromagnetic radiation is quantized by treating each mode as an equivalent harmonic oscillator. From this analogy, every mode of the field must have hf/2 as its average minimum energy. That is a tiny amount of energy, but the number of modes is enormous, and indeed increases as the square of the frequency. The product of the tiny energy per mode times the huge spatial density of modes yields a very high theoretical energy density per cubic centimeter.

From this line of reasoning, quantum physics predicts that all of space must be filled with electromagnetic zero-point fluctuations (also called the zero-point field) creating a universal sea of zero-point energy. The density of this energy depends critically on where in frequency the zero-point fluctuations cease. Since space itself is thought to break up into a kind of quantum foam at a tiny distance scale called the Planck scale (10^{-33} cm), it is argued that the zero point fluctuations must cease at a corresponding Planck frequency (10^{43} Hz). If that is the case, the zero-point energy density would be 110 orders of magnitude greater than the radiant energy at the center of the Sun.

Connection To Inertia And Gravitation

When a passenger in an airplane feels pushed against his seat as the airplane accelerates down the runway, or when a driver feels pushed to the left when her car makes a sharp turn to the right, what is doing the pushing? Since the time of Newton, this has been attributed to an innate property of matter called inertia. In 1994 a process was discovered whereby the zero-point fluctuations could be the source of the push one feels when changing speed or direction, both being forms of acceleration. The zero-point fluctuations could be the underlying cause of inertia. If that is the case, then we are actually sensing the zero-point energy with every move we make (see origin of inertia http://www.calphysics.org/inertia.html).

The principle of equivalence would require an analogous connection for gravitation. Einstein's general relativity successfully accounts for the motions of freely-falling objects on geodesics (the "shortest" distance between two points in curved spacetime), but does not provide a mechanism for generating a gravitational force for objects when they are forced to deviate from geodesic tracks. It has been found that an object undergoing acceleration or one held fixed in a gravitational field would experience the same kind of asymmetric pattern in the zero-point field giving rise to such a reaction force. The weight you measure on a scale would therefore be due to zero-point energy (see http://www.calphysics.org/gravitation.html).

The possibility that electromagnetic zero-point energy may be involved in the production of inertial and gravitational forces opens the possibility that both inertia and gravitation might someday be controlled and manipulated. This could have a profound impact on propulsion and space travel.

PRIMARY ARTICLES

(see Scientific Articles http://www.calphysics.org/sci_articles.html for additional articles)

1) Gravity and the Quantum Vacuum Inertia Hypothesis http://www.calphysics.org/articles/gravity_arxiv.pdf, Alfonso Rueda & Bernard Haisch, Annalen der Physik, Vol. 14, No. 8, 479-498 (2005).
2) Review of Experimental Concepts for Studying the Quantum Vacuum Fields http://www.calphysics.org/articles/Davis_STAIF06.pdf, E. W. Davis, V. L. Teofilo, B. Haisch, H. E. Puthoff, L. J. Nickisch, A. Rueda and D. C. Cole, Space Technology and Applications International Forum (STAIF 2006), p. 1390 (2006).
3) Analysis of Orbital Decay Time for the Classical Hydrogen Atom Interacting with Circularly Polarized Electromagnetic Radiation http://www.calphysics.org/articles/ColeHydrogenPRE.pdf, Daniel C. Cole & Yi Zou, Physical Review E, 69, 016601, (2004).
4) Inertial mass and the quantum vacuum fields http://xxx.arxiv.org/abs/gr-qc/0009036, Bernard Haisch, Alfonso Rueda & York Dobyns, Annalen der Physik, Vol. 10, No. 5, 393-414 (2001).

5) Stochastic nonrelativistic approach to gravity as originating from vacuum zero-point field van der Waals forces http://www.calphysics.org/articles/Cole_Rueda_Danley.pdf, Daniel C. Cole, Alfonso Rueda, Konn Danley, Physical Review A, 63, 054101, (2001).
6) The Case for Inertia as a Vacuum Effect: a Reply to Woodward & Mahood http://xxx.arxiv.org/abs/gr-qc/0002069, Y. Dobyns, A. Rueda & B.Haisch, Foundations of Physics, Vol. 30, No. 1, 59 (2000).
7) On the relation between a zero-point-field-induced inertial effect and the Einstein-de Broglie formula http://xxx.arxiv.org/abs/gr-qc/9906084, B. Haisch & A. Rueda, Physics Letters A, 268, 224, (2000).
8) Contribution to inertial mass by reaction of the vacuum to accelerated motion http://xxx.arxiv.org/abs/physics/9802030, A. Rueda & B. Haisch, Foundations of Physics, Vol. 28, No. 7, pp. 1057-1108 (1998).
9) Inertial mass as reaction of the vacuum to acccelerated motion http://xxx.arxiv.org/abs/physics/9802031, A. Rueda & B. Haisch, Physics Letters A, vol. 240, No. 3, pp. 115-126, (1998).
10) Reply to Michel's "Comment on Zero-Point Fluctuations and the Cosmological Constant" http://www.calphysics.org/articles/zpf_apj.pdf, B. Haisch & A. Rueda, Astrophysical Journal, 488, 563, (1997).
11) Quantum and classical statistics of the electromagnetic zero-point-field http://www.calphysics.org/articles/PRA96.pdf, M. Ibison & B. Haisch, Physical Review A, 54, pp. 2737-2744, (1996).
12) Vacuum Zero-Point Field Pressure Instability in Astrophysical Plasmas and the Formation of Cosmic Voids <http://adsbit.harvard.edu/cgi-bin/nph-iarticle_query?1995ApJ%2E%2E%2E445%2E%2E%2E

%2E7R&nosetcookie=1Vacuum Zero-Point-Field Pressure>, A. Rueda, B. Haisch & D.C. Cole, Astrophysical Journal, Vol. 445, pp. 7-16 (1995).

13) Inertia as a zero-point-field Lorentz force http://www.calphysics.org/articles/PRA94.pdf, B. Haisch, A. Rueda & H.E. Puthoff, Physical Review A, Vol. 49, No. 2, pp. 678-694 (1994).

ZPE in '12?

Aviation Week & Space Technology
00052175, 3/1/2004,Vol. 160, Issue 9
From http://weblinks1.epnet.com/citation.asp?tb=1&_ua=bo

In trying to predict when a scientific breakthrough might unlock ***zero point energy*** (ZPE) as a space transportation power source, a few scientists suggest looking for clues in historical cycles.

One of the more enticing is the Kondratieff interval, which was defined by Nikolai Kondratieff in 1924. Often cited in economic studies, the roughly 55-year cycle can be found in a variety of human-event patterns. John E. Allen, a longtime aerospace researcher and consultant for BAE Systems, found that the Kondratieff cycle shows up in key milestones leading to spaceflight (see chart). If the cycle holds true, then mankind is due for another breakthrough in about 2012 — which will be 55 years after the launch of Russia's Sputnik, mankind's first satellite, and 109 years after the Wright brothers' first flight.

Hal E. Puthoff, one of several scientists who has spent years trying to "break the code" that would release the tremendous potential of ZPE, quipped, "It's always darkest just before it's pitch black (see p. 50). The most frustrating period is when you know [the answer] is close, but you're not there yet. Certainly, that's where we are now. The fact that major aerospace companies are getting interested in [ZPE] will definitely accelerate the process. But there's no way to predict how long that'll take."

He and other ZPE researchers might look not only to the Kondratieff interval for encouragement, but also to Wilbur Wright's recollection in 1908: "In 1901, I confess that I said to my brother, Orville, that man would not fly for 50 years. Two years later, we made flights. This demonstration of my impotence as a prophet gave me such shock that ever more I have distrusted myself and avoided all predictions."

Energy Unlimited

Henry Bortman, *New Scientist* magazine, 22/1/2000

Empty space is seething with huge quantities of energy-- if only we could tap it. Henry Bortman reports on a micromachine designed to do just that.

"I am busy just now again on electro-magnetism, and think I have got hold of a good thing, but can't say. It may be a weed instead of a fish that, after all my labor, I may at last pull up."

The writer was Michael Faraday and his catch turned out to be a very big fish indeed. Faraday's work on the relationship between electricity and magnetism was among the most important research of the 19th century, but his writings give a unique insight into the worries facing a scientist working at the edge of human knowledge. You could say Jordan Maclay is in a similar position.

Last year, Maclay secured funding from NASA to study the energy of a vacuum. His research is part of the Breakthrough Propulsion Physics programme based at the NASA Glenn Research Center in Cleveland, Ohio. The goal of the programme is to find new methods of propulsion that could power spacecraft. His plan is to build a tiny machine that will measure this vacuum energy and the forces it can produce. If things go well, Maclay could land a fish of gargantuan proportions. He hopes to find a way of exploiting these forces to do something useful such as drive a miniature piston, heat water, or even power a spacecraft. Next

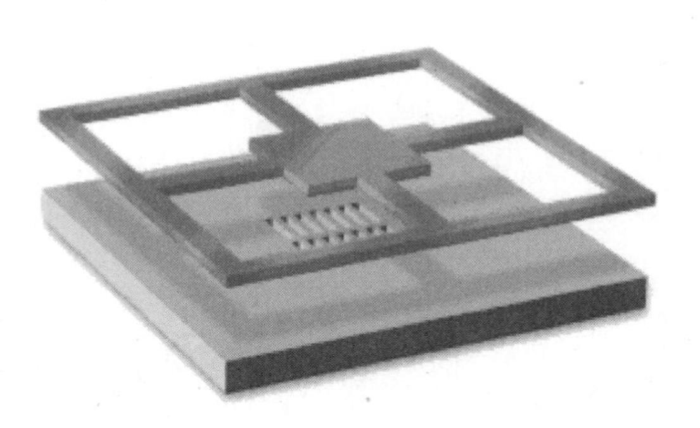

week, he will outline his plans at a meeting of the Space Technology and Applications International Forum (STAIF) in Albuquerque, New Mexico.

Most people assume that the vacuum is empty. But according to quantum electrodynamics, the theory that describes the behaviour of the Universe at the very small scale, nothing could be further from the truth. The vacuum is actually seething with electromagnetic energy called zero-point energy and it's this that Maclay hopes to tap. The "zero" in zero-point refers to the fact that if you were to cool the Universe to absolute zero, its lowest possible energy state, some energy would remain. Actually, rather a lot of energy. Physicists disagree over just how much, but Maclay has calculated that a region of the vacuum the size of a proton could contain as much energy as all the matter in the entire Universe.

In 1948, a Dutch physicist called Hendrick Casimir proposed a scheme to test for the presence of this energy. In theory, vacuum energy takes the form of particles that are constantly forming and disappearing on a tiny scale. Normally, the vacuum is filled with particles of almost any wavelength, but Casimir argued that if you were to place two thin uncharged metal plates very close together, longer wavelengths would be excluded. The extra waves outside the plates would then generate a force that tended to push them together, and the closer the plates were together, the stronger the attraction would be. In 1996, physicists measured the so-called Casimir effect for the first time.

Maclay, a former professor of electrical engineering at the University of Illinois in Chicago, wants to go further and has formed a company called Quantum Fields in Richland Center, Wisconsin, to develop his ideas. He and others have calculated that the Casimir effect can produce repulsive forces as well as attractive ones. His analysis has focused not on metal plates but on tiny metal boxes, roughly 1 micrometre or less on each side, which he refers to as cavities (see diagram).

It turns out that the Casimir force, and its direction, depend on the shape of the cavity. "If you have a cavity the shape of a pizza box, the pressure on the two large sides of the box pushes them together, but the force on the narrow sides pushes them apart," he says.

The cavity Maclay finds most intriguing is long and thin, like the box a tube of toothpaste comes in, and about the size of an *Escherichia coli* bacterium. What's significant about this cavity is that one of its long sides is at perfect equilibrium: the inward and outward vacuum pressures are exactly equal. But it's a tenuous equilibrium. And that's what makes it interesting.

Maclay plans to build a box in which the side at equilibrium--call it the lid--is free to move. If the lid moves inward slightly from the equilibrium point, the vacuum pressure inside the cavity goes down, and the lid is drawn farther in. If it moves outward the reverse happens and the lid is pushed away. The distances involved are tiny--less than 100 nanometres. The lid will be attached to a microscopic spring. So when the lid moves, the spring will be stretched or compressed and will tend to return to its original position. Maclay is hoping that by carefully balancing the vacuum pressure of the cavity and the elasticity of the spring, and by giving the lid just the right initial impulse, he can create a tiny oscillator driven by Casimir forces.

That's the ideal scenario, at any rate. But it's easy to make things look good on paper. "From a theoretical viewpoint," says Maclay, "all kinds of things oscillate. But in the real world? Well, that's what we have to look at." Maclay plans to attack the problem in stages. Repulsive Casimir forces have never been measured so his first task will be to find out if he can even do this. Next he'll measure the inward and outward forces at the surfaces of cavities with different shapes, to see if they match predictions. And if all that goes well, *he'll be ready to build a resonating cavity.*

The job of building the experimental setup falls to Rod Clark, a former nuclear engineer and president of MEMS Optical, a technology company based in Huntsville, Alabama, that manufactures microelectromechanical devices (MEMs). To build Maclay's cavities out of silicon, Clark hopes to use a combination of traditional lithographic etching and deposition techniques--the same techniques used to make integrated circuits.

Clark is confident that he can produce the necessary structures. But he's also well aware of the challenges, the first of which is size. Maclay's specifications are at the limits of today's fabrication technology, says Clark. "We want to make it small in order to make the forces large. But we can't make it so small that we can't fabricate it."

Maclay and Clark's current plan is to make an array of hundreds of topless cavities on a substrate, and then create a single lid that fits over the entire array. The lid will be suspended on springs above the array, which will be moved toward the lid in tiny steps. Initially the lid should remain still, but when the cavities get close enough, the difference in vacuum pressure should cause it to move and possibly even to oscillate. By peering across the surface of the lid through a microscope, it will be possible to measure its displacement with great precision.

Neither Maclay nor Clark expects quick results. Over the course of the three-year period covered by NASA's grant, they hope to build three generations of devices. Still, Maclay is already dreaming of various types of "Casimir machines" that might be possible if his experiments prove successful. Microscopic vacuum-drive levers, pulleys and pistons come to mind, for example. Or perhaps a machine that contains cavities that generate different vacuum pressures and exploits that difference in much the same way that a heat engine exploits differences in temperature. "What we're looking at now are very simple things that ultimately will serve as components of more complicated systems," he says. "We've gotta kind of mess around to see what they can do."

How does Maclay rate his chances of success? "If I thought the chance was zero," he says, "I wouldn't spend my time on it. I'm convinced we'll find some interesting things. Exactly what and what its utility will be, I don't know."

Marc Millis, who heads NASA's Breakthrough Propulsion Physics programme, is also philosophical about the prospects. He'd be thrilled, of course, if Maclay handed him the keys to an interstellar propulsion drive. "I would be very surprised if there wasn't a potential breakthrough of some kind," he says of the entire propulsion project. If Maclay is able to reel in his fish, he could have a phenomenon to rival Faraday's in importance. And as Faraday once said: "Nothing is too wonderful to be true if it be consistent with the laws of nature."

Further reading:

- Jordan Maclay's website is at www.quantumfields.com
- For more information on the Casimir effect see math.ucr.edu/home/baez/physics/casimir.html
- The Breakthrough Propulsion Programme is at www.grc.nasa.gov/WWW/bpp/

Quantum Vacuum Fluctuations: A New Rosetta Stone of Physics?

Dr. H. E. Puthoff
Institute for Advanced Studies
1301 Capital Of Texas Highway S., Suite B 121
Austin, Texas 78746
(512) 328-5751

In a recent article in the popular press (*The Economist*, January 7, 1989, pp. 71-74) it was noted how many of this century's new technologies depend on the Alice-in-Wonderland physics of quantum mechanics, with all of its seeming absurdities. For starters, one begins with the observation that classical physics tells us that atoms, which can be likened to a miniature solar system with electron planets orbiting a nuclear sun, should not exist. The circling electrons should radiate away their energy like microscopic radio antennas and spiral into the nucleus. But atoms do exist, and multitudinous other phenomena which don't obey the rules do occur. To resolve this cognitive dissonance physicists introduced quantum mechanics, which is essentially a set of mathematical rules to describe what in fact does happen. But when we re-ask the question, "why didn't the electron radiate away its energy?" the answer is, basically, "well, in quantum theory it doesn't." It's at this point that not only the layman but some physicists can begin to feel that someone's not playing fair. I say only some physicists because the majority of working physicists are content simply to use quantum rules that work, that describe (if only statistically) what will happen in a given experiment under certain conditions.

These are the so-called "logical positivists" who, in a philosophical sense, are like the news reporter whose only interest is the bottom line. There are nevertheless individuals here and there who still want to know why the electron didn't radiate, why Einstein's equations are in this form and not another, where does the ubiquitous zero- point energy that fills even empty space come from, why quantum theory, and perhaps the biggest question of all, how did the universe get started anyway? Surprisingly enough, there may be answers to these seemingly unanswerable meta-level questions. Perhaps even more surprising, they seem to be emerging, as a recent book title put it, from "Something called Nothing" (**1**), or to put it more correctly, from empty space, the vacuum, the **void**. To comprehend the significance of this statement, we will have to take a detour into the phenomenon of fluctuations with which quantum theory abounds, including the fluctuations of empty space itself. Before the advent of quantum theory, physics taught that any simple oscillator such as a pendulum, when excited, would eventually come to rest if not continuously energized by some outside force such as a spring. This is because of friction losses in the system.

After it was recognized that quantum theory was a more accurate representation of nature, one of the findings of quantum theory was that such an oscillator would in fact not come to total rest but rather would continue to "jiggle" randomly about its resting point with a small amount of energy always present, the so-called "zero-point energy." Although it may not be observable to the eye on your grandfather clock because it is so minute, it is nonetheless very real, and in many physical systems has important consequences. One example is the presence of a certain amount of "noise" in a microwave receiver that can never be gotten rid of, no matter how perfect the technology. This is an example which shows that not only physical devices such as pendulums have this property of incessant fluctuation, but also fields, such as electromagnetic fields (radio waves, microwaves, light, X-rays, etc.).

As it turns out, even though the zero-point energy in any particular mode of an electromagnetic field is minute, there are so many possible modes of propagation (frequencies, directions) in open space, the zero-point energy summed up over all possible modes is quite enormous; in fact, greater than, for example, nuclear energy densities. And this in all of so-called "empty" space around us. Let us concentrate on the effects of such electromagnetic zero-point fluctuations. With such large values, it might seem that the effects of electromagnetic zero-point energy should be quite obvious, but this is not the case because of its extremely uniform density. Just as a vase standing in a room is not likely to fall over spontaneously, so a vase bombarded uniformly on all sides by millions of ping pong balls would not do likewise because of the balanced conditions of the uniform bombardment. The only evidence of such a barrage might be minute jiggling of the vase, and similar mechanisms are thought to be involved in the quantum jiggle of zero-point motion.

However, there are certain conditions in which the uniformity of the background electromagnetic zero-point energy is slightly disturbed and leads to physical effects. One is the slight perturbation of the lines seen from transitions between atomic states known as the Lamb Shift (**2**), named after its discoverer, Willis Lamb. Another, also named for its discoverer, is the Casimir Effect, a unique attractive quantum force between closely-spaced metal plates. An elegant analysis by Milonni et. al. at Los Angeles National Laboratory (**3**) shows the Casimir force to be due to radiation pressure from the background electromagnetic zero-point energy which has become unbalanced due to the presence of the plates, and which results in the plates being pushed together. From this it would seem that it might be possible to extract electrical energy from the vacuum, and indeed the possibility of doing so (at least in principle) has been shown in a paper of that same name by Robert Forward (**4**) at Hughes Research Laboratories in Malibu, California.

What does this have to do with our basic questions? Let's start with the question as top why the electron in a simple hydrogen atom doesn't radiate as it circles the proton in its stable ground state atomic orbit. This issue has been re-addressed in a recent paper by the author, this time taking into account what has been learned over the years about the effects of zero-point energy. (**5**) There it is shown that the electron can be seen as continually radiating away its energy as predicted by classical theory, but simultaneously absorbing a compensating amount of energy from the ever-present sea of zero-point energy in which the atom is immersed, and an assumed equilibrium between these two processes leads to the correct values for the parameters known to define the ground-state orbit. Thus the ground-state orbit is set by a dynamic equilibrium in which collapse of the state is prevented by the presence of the zero-point energy. The significance of this observation is that **the very stability of matter itself appears to depend on the presence of the underlying sea of electromagnetic zero-point energy.**

With regard to the gravitational attraction that is described so well by Einstein's theory, its fundamental nature is still not well understood. Whether addressed simply in terms of Newton's Law, or with the full rigor of general relativity, gravitational theory is basically descriptive in nature, without revealing the underlying dynamics for that description. As a result, attempts to unify gravity with the other forces (electromagnetic, strong and weak nuclear forces) or to develop a quantum theory of gravity have foundered again and again on difficulties that can be traced back to a lack of understanding at a fundamental level. To rectify these difficulties, theorists by and large have resorted to ever-increasing levels of mathematical sophistication and abstraction, as in the recent development of supergravity and superstring theories. Taking a completely different tack when addressing these difficulties in the sixties, the well-known Russian physicist Andrei Sakharov put forward the somewhat radical hypothesis that gravitation might not be a fundamental interaction at all, but rather a secondary or residual effect associated with other (non- gravitational) fields. (**6**)

Specifically, Sakharov suggested that gravity might be an induced effect brought about by changes in the zero-point energy of the vacuum, due to the presence of matter. If correct, gravity would then be understood as a variation on the Casimir theme, in which background zero-point-energy pressures were again responsible. Although Sakharov did not develop the concept much further, he did outline certain criteria such a theory would have to meet such as predicting the value of the gravitational constant G in terms of zero-point-energy parameters. The approach to gravity outlined by Sakharov has recently been addressed in detail, and with positive results, again by the author. (**7**) The gravitational interaction is shown to begin with the fact that a particle situated in the sea of electromagnetic zero-point fluctuations develops a "jitter" motion, or *ZITTERBEWEGUNG* as it is called. When there are two or more particles they are each influenced not only by the fluctuating background field, but also by the fields generated by the other particles, all similarly undergoing ZITTERBEWEGUNG motion, and the inter-particle coupling due to these fields results in the attractive gravitational force.

Gravity can thus be understood as a kind of long-range Casimir force. Because of its electromagnetic underpinning, gravitational theory in this form constitutes what is known in the literature as an "already-unified" theory. The major benefit of the new approach is that it provides a basis for understanding various characteristics of the gravitational interaction hitherto unexplained. These include the relative weakness of the gravitational force under ordinary circumstances (shown to be due to the fact that the coupling constant G depends inversely on the large value of the high-frequency cutoff of the zero-point-fluctuation spectrum); the existence of positive but not negative mass (traceable to a positive-only kinetic-energy basis for the mass parameter); and the fact that gravity cannot be shielded (a consequence of the fact that quantum zero-point-fluctuation "noise" in general cannot be shielded, a factor which in other contexts sets a lower limit on the delectability of electromagnetic signals).

As to where the ubiquitous electromagnetic zero-point energy comes from, historically there have been two schools of thought: existence by fiat as part of the boundary conditions of the universe, or generation by the (quantum-fluctuation) motion of charged particles that constitute matter. A straightforward calculation of the latter possibility has recently been carried out by the author. (**8**) It was assumed that zero-point fields drive particle motion, and that the sum of particle motions throughout the universe in turn generate the zero-point fields, in the form of a self-regenerating cosmological feedback cycle not unlike a cat chasing its own tail. This self-consistent approach yielded the known zero-point field distribution, thus indicating a dynamic-generation process for the zero-point fields. Now as to the question of why quantum theory. Although knowledge of zero-point fields emerged from quantum physics as that subject matured, Professor Timothy Boyer at City College in New York took a contrary view.

He began asking in the late sixties what would happen if we took classical physics as it was and introduced a background of random, classical fluctuating fields of the zero-point spectral distribution type. Could such an all-classical model reproduce quantum theory in its entirety, and might this possibility have been overlooked by the founders of quantum theory who were not aware of the existence of such a fluctuating background field? (First, it is clear from the previously-mentioned cosmological calculation that such a field distribution would reproduce itself on a continuing dynamic basis.) Boyer began by tackling the problems that led to the introduction of quantum theory in the first place, such as the blackbody radiation curve and the photoelectric effect. One by one the known quantum results were reproduced by this upstart neoclassical approach, now generally referred to as Stochastic Electrodynamics (SED) (**9**), as contrasted to quantum electrodynamics (QED). Indeed, Milonni at Los Alamos noted in a review of the Boyer work that had physicists in 1900 thought of taking this route, they would probably have been more comfortable with this classical approach than with Planck's hypothesis of the quantum, and one can only speculate as to the direction that physics would have taken then.

The list of topics successfully analyzed within the SED formulation (i.e., yielding precise quantitative agreement with QED treatments) has now been extended to include the harmonic oscillator, Casimir and Van der Waals forces and the thermal effects of acceleration through the vacuum, to name a few. Out of this work emerged the reasons for such phenomena as the uncertainty principle, the incessant fluctuation of particle motion, the existence of Van der Waals forces even at zero temperature, and so forth, all shown to be due to the influence of the unceasing activity of the random background fields. There are also some notable failures in SED, such as transparent derivation of something as simple as Schrodinger's equation, which turns out as yet to be an intractable problem. Therefore, it is unlikely that quantum theory as we have come to know it and love it will be entirely replaced by a refurbished classical theory in the near future. Nonetheless, the successes to date of the SED approach, by its highlighting of the role of background zero-point-fluctuations, means that when the final chapter is written on quantum theory, field fluctuations in empty space will be accorded an honored position.

And now to the preeminent question of all, where did the Universe come from? Or, in modern terminology, what started the Big Bang? Could quantum fluctuations of empty space have something to do with this also? Well, Prof. Edward Tryon of Hunter College of the City University of New York thought so when he proposed in 1973 that our Universe may have originated as a fluctuation of the vacuum on a large scale, as "simply one of those things which happen from time to time." (**10**) This idea was later refined and updated within the context of inflationary cosmology by Alexander Vilenkin of Tufts University, who proposed that the universe is created by quantum tunneling from literally nothing into the something we call our universe. (**11**) Although highly speculative, these types of models indicate once again that physicists find themselves turning again and again to the Void (and the fluctuations thereof) for their answers.

Those with a practical bent of mind may be left with yet one more unanswered question. Can this emerging Rosetta Stone of physics be used to translate such lofty insights into mundane application? Could the engineer of the future specialize in "vacuum engineering?" Could the energy crisis be solved by harnessing the energies of the zero-point sea? After all, since the basic zero-point energy form is highly random in nature, and tending towards self-cancellation, if a way could be found to bring order out of chaos, the, because of the highly energetic nature of the vacuum fluctuations, relatively large effects could in principle be produced. Given our relative ignorance at this point, we must fall back on a quote given by Podolny (**12**) when contemplating this same issue.

> "It would be just as presumptuous to deny the feasibility of useful application as it would be irresponsible to guarantee such application." Only the future can reveal the ultimate use to which Mankind will put this remaining Fire of the Gods, the quantum fluctuations of empty space.

REFERENCES

1) R. Podolny, "Something Called Nothing" (Mir Publ., Moscow,1986)

2) W. E. Lamb, Jr., and R. C. Retherford, "Fine Structure of the Hydrogen Atom by a Microwave Method," Phys. Rev. 72, 241 (1947)

3) P. W. Milonni, R. J. Cook and M. E. Goggin, "Radiation Pressure from the Vacuum : Physical Interpretation of the Casimir Force," Phys. Rev. A 38, 1621 (1988)

4) R. L. Forward, "Extracting Electrical Energy from the Vacuum by Cohesion of Charged Foliated Conductors," Phys. Rev. B 30, 1700 (1984)

5) H. E. Puthoff, "Ground State of Hydrogen as a Zero-Point

Fluctuation-Determined State," Phys. Rev. D 35, 3266 (1987) See also science news article, "Why Atoms Don't Collapse," in New Scientist, p. 26 (9 July 1987)

6) A. D. Sakharov, "Vacuum Quantum Fluctuations in Curved Space and the Theory of Gravitation, Dokl. Akad. Nauk. SSSR (Sov. Phys. - Dokl. 12, 1040 (1968). See also discussion in C. W. Misner, K. S. Thorne and J. A. Wheeler, Gravitation (Freeman, San Francisco,1973), p. 426

7) H. E. Puthoff, "Gravity as a Zero-Point Fluctuation Force," Phys. Rev. A 39, 2333 (1989)

8) H. E. Puthoff, "Source of Vacuum Electromagnetic Zero-Point Energy," subm. to Phys. Rev. A, (March 1989)

9) See review of SED by T. H. Boyer, "A Brief Survey of Stochastic Electrodynamics," in Foundations of Radiation Theory and Quantum Electrodynamics, edited by A. O. Barut (Plenum, New York, 1980) See also the very readable account "The Classical Vacuum," in Scientific American, p. 70 (August 1985)

10) E. P. Tryon, "Is the Universe a Vacuum Fluctuation?" Nature 246, 396 (1973)

11) A. Vilenkin, "Creation of Universes from Nothing," Phys. Lett. 117B, 25 (1982)

12) R. Podolny, Ref. 1, p. 211

Reprinted from http://www.clas.ufl.edu/anthro/ZPE.html

Institute of Physics Publishing Journal of Physics: Conference Series 31 (2006) 123–130
doi:10.1088/1742-6596/31/1/021 The Third 21COE Symposium: Astrophysics as Interdisciplinary Science

Laboratory tests on dark energy

Christian Beck

School of Mathematical Sciences, Queen Mary, University of London, Mile End Road, London E1 4NS, UK

E-mail: c.beck@qmul.ac.uk

Abstract. The physical nature of the currently observed dark energy in the universe is completely unclear, and many different theoretical models co-exist. Nevertheless, if dark energy is produced by vacuum fluctuations then there is a chance to probe some of its properties by simple laboratory tests based on Josephson junctions. These electronic devices can be used to perform 'vacuum fluctuation spectroscopy', by directly measuring a noise spectrum induced by vacuum fluctuations. One would expect to see a cutoff near 1.7 THz in the measured power spectrum, provided the new physics underlying dark energy couples to electric charge. The effect exploited by the Josephson junction is a subtle nonlinear mixing effect and has nothing to do with the Casimir effect or other effects based on van der Waals forces. A Josephson experiment of the suggested type will now be built, and we should know the result within the next 3 years.

1. Introduction—what is dark energy?

It would be nice to start this paper with a clear definition of what dark energy is and what it is good for. Unfortunately, the answer to this question is completely unclear at the moment. What is clear is that various astronomical observations [1, 2] (supernovae, CMB fluctuations, large-scale structure) provide rather convincing evidence that around 73% of the energy contents of the universe is a rather homogeneous form of energy, so-called 'dark energy'. It behaves similar to a cosmological constant and currently causes the universe to accelerate its expansion. Dark energy may just be vacuum energy (with an equation of state $w = p/\rho = -1$, where p denotes the pressure and ρ the energy density). In that case its energy density ρ is constant and does not change with the expansion of the universe. Or, w may be just close to -1, in which case the dark energy density evolves dynamically and changes with the expansion of the universe. The remaining contents of the current universe is about 23% dark matter and 4% ordinary matter. With 96% of the universe being unknown stuff, there is enough room (and, indeed, the need) for new theories. It seems that in order to construct a convincing theory of dark energy that explains why it is there and what role it plays in the universe one has to be open-minded to new physics.

A large number of different theoretical models exist for dark energy, but an entirely convincing theoretical breakthrough has not yet been achieved. Popular models are based on quintessence fields, phantom fields, quintom fields, Born-Infeld quantum condensates, the Chaplygin gas, fields with nonstandard kinetic terms, to name just a few (see e.g. [3, 4, 5, 6] for reviews). All of these approaches contain 'new physics' in one way or another, though at different levels. However, it is clear that the number of possible dark energy models that are based on new physics is infinite, and in that sense many other models can be considered as well. Only experiments will ultimately be able to confirm or refute the various theoretical constructs.

2. Dark energy from vacuum fluctuations

A priori the simplest explanation for dark energy would be to associate it with vacuum fluctuations that are allowed due to the uncertainty relation. From quantum field theory it is well known that virtual momentum fluctuations of a particle of mass m and spin j formally produce vacuum energy given by

$$\rho_{vac} = \frac{1}{2}(-1)^{2j}(2j+1)\int_{-\infty}^{+\infty} \frac{d^3k}{(2\pi)^3}\sqrt{\mathbf{k}^2+m^2} \tag{1}$$

in units where $\hbar = c = 1$. Here $\mathbf{k}$ represents the momentum and the energy is given by $E = \sqrt{\mathbf{k}^2+m^2}$. Unfortunately, the integral is divergent: It formally yields infinite vacuum energy density. Hence one has to find excuses why this type of vacuum energy is not observable, or why only a very small fraction of it survives.

A typical type of argument is that vacuum energy as given by eq. (1) is not gravitationally active, as long as the theory is not coupled to gravity. This is more a belief rather than a proved statement, and the statement looks a bit fragile: Most particles in the standard model of electroweak and strong interactions do have mass, so they know what gravity is and hence it is not clear why their vacuum energy should not be creating a gravitational effect if their mass does. Another argument is that the vacuum energy might be cancelled by some unknown symmetry, for example some type of supersymmetry. For supersymmetric models, fermions with the same mass as bosons generate vacuum energies of opposite sign, so all vacuum energies add up to zero. Unfortunately, we know that we currently live in a universe where supersymmetry is broken: Nobody has ever observed bosonic electrons, having the same mass as ordinary electrons. So this idea does not work either. We see that in both cases the problem why we can't honestly get rid of the vacuum energy comes from the masses of the particles, and there is no theory of masses so far. In fact the Higgs mechanism, which is supposed to create the particle masses in the standard model of electroweak and strong interactions, creates an additional (unacceptable) amount of vacuum energy, due to symmetry breaking. So the only thing that we can say for sure is that the problem of masses and vacuum energy in the current universe is not yet fully understood.

The next, more pragmatic, step is then to introduce an upper cutoff for the momentum $|\mathbf{k}|$ in the integral (1). This leads to finite vacuum energy. But typically, one would expect this cutoff to be given by the Planck mass m_{Pl}, since we expect quantum field theory to be replaced by a more sophisticated theory at this scale. Unfortunately, this gives a vacuum energy density of the order m_{Pl}^4—too large by a factor 10^{120} as compared to current measurements of dark energy density. This is the famous cosmological constant problem. The presently observed dark energy density is more something of the order m_ν^4, where m_ν is a typical neutrino mass scale.

To explain the currently observed small dark energy density in the universe, it seems one has to be open-minded to new physics. This new physics can enter at various points. If vacuum fluctuations (of whatever type) create dark energy, we basically have two possibilities: Either, the dark energy is created by vacuum fluctuations of ordinary particles of the standard model (e.g. photons). The corresponding vacuum energy is given by the integral (1), and the new physics should then explain why there is a cancellation of the vacuum energy if the momentum $|\mathbf{k}|$ exceeds something of the order m_ν. Or, there may be new types of vacuum fluctuations created by a new type of dynamics underlying the cosmological constant which intrinsically has a cutoff at around m_ν. A priori, there is no reason why this new dynamics of vacuum fluctuations should not couple to electric charge. In fact, ultimately there is the need to unify the quantum field theory underlying the standard model with gravity, so the missing piece in this jigsaw should couple to both. If the new dynamics of vacuum fluctuations (or the new cancellation process of vacuum energy) couples to electric charge, then there is a chance to see effects of this in laboratory experiments on the earth [7], as we shall work out in the following.

3. Vacuum fluctuation spectroscopy with Josephson junctions

It is well known that vacuum fluctuations produce measurable noise in dissipative systems (see, e.g., [8] and references therein). This noise has been experimentally verified in many experiments. What is the deeper reason for the occurence of quantum noise, e.g., in resistors? The basic principle is quite easy to understand. Consider a dissipative system and two canonically conjugated variables, say x (position) and p (momentum). If there are no external forces, then in the long-term run the system will classically reach the stable fixed point $p = 0$, since all kinetic energy is dissipated. But quantum mechanically, a state with $p = 0$ would contradict the uncertainty principle $\Delta x \Delta p = O(\hbar)$. Hence there must be noise in the resistor that keeps the momentum going. This noise acts as a fluctuating driving force and makes sure that the momentum does not take on a fixed value, so that the uncertainty principle is satisfied at all times.

Quantum noise is a dominant effect if the temperature is small. It can be directly measured in experiments. Maybe one of the most impressive experiments in this direction is the one done by Koch et al. in the early eighties, based on resitively shunted Josephson junctions [9, 10]. A Josephson junction consists of two superconductors with an insulator sandwiched inbetween. The behaviour of such a device can be modeled by the following stochastic differential equation

$$\frac{\hbar C}{2e}\ddot{\delta} + \frac{\hbar}{2eR}\dot{\delta} + I_0 \sin\delta = I + I_N. \qquad (2)$$

Here δ is the phase difference across the junction, R is the shunt resistor, C the capacitance of the junction, I is the mean current, I_0 the noise-free critical current, and I_N is the noise current. One can think of eq. (2) as formally describing a particle that moves in a tilted periodic potential. The noise current I_N produces some perturbations to the angle of the tilt. There are two main sources of noise: Thermal fluctuations and quantum fluctuations.

Koch et al. [10] measured the noise spectrum at low temperatures using four different Josephson devices. They experimentally verified the following form of the spectrum up to frequencies of $6 \cdot 10^{11} Hz$:

$$\begin{aligned} S(\nu) &= \frac{2h\nu}{R}\coth\left(\frac{h\nu}{2kT}\right) \\ &= \frac{4h\nu}{R}\left(\frac{1}{2} + \frac{1}{\exp(h\nu/kT) - 1}\right). \end{aligned} \qquad (3)$$

The first term in Equation (3) grows linear with the frequency, as expected for the ground state of a quantum mechanical oscillator. This term represents quantum noise and is induced by zero-point fluctuations. The second term is ordinary Bose-Einstein statistics and corresponds to thermal noise.

What is remarkable with the experiment of Koch et al. is the fact that the quantum fluctuations produce a directly measurable spectrum—many theoretical physicists thought (and some still think) that this is not possible. The experimentally measured data of Koch et al. are shown in Fig. 1. In fact, noise induced by quantum fluctuations plays an important role in any resistor if the temperature is small enough. The Josephson junction serves as a useful technical tool in this context: Due to a nonlinear mixing effect in the junction, one has the experimental possibility to obtain measurements of very high frequency noise. For recent theoretical work on the quantum noise theory of Josephson junctions, see [8, 11].

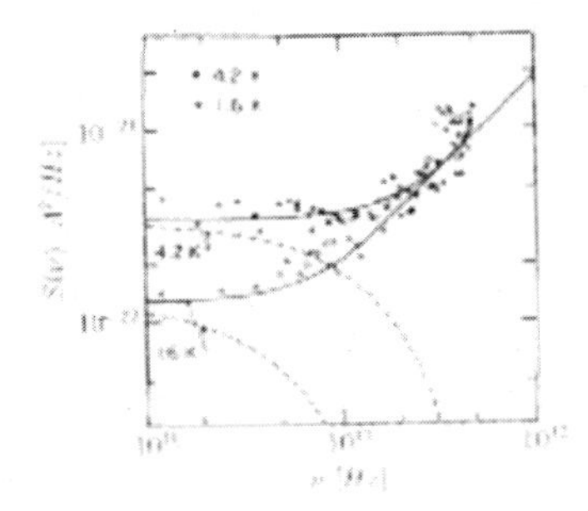

Figure 1. Spectral density of current noise as measured in Koch et al.'s experiment [10] for two different temperatures. The solid line takes into account the zero-point term and is given by eq. (3), whereas the dashed line is given by a purely thermal noise spectrum, $(4h\nu/R)(exp(h\nu/kT)-1)^{-1}$.

The linear term in the spectrum is induced by zero-point fluctuations and thus a consequence of the uncertainty relation. Contrary to this, Jetzer and Straumann [12] have recently expressed the view that the linear growth in the measured spectrum is produced by van der Waals forces and that it has nothing to do with zero-point fluctuations. But their view seems to be at variance with the standard view of almost all experts in the field of quantum noise theory (see [8] and references therein). The standard view is that the linear term in the measured spectrum is indeed a fingerprint of zero-point fluctuations of the harmonic oscillators which model the microscopic structure of the resistive element. Van der Waals forces, as well as the Casimir effect, are different effects that are relevant for other types of experimental situations (see, e.g. [13]), not for the noise in Josephson junctions.

Since zero-point fluctuations produce experimentally measurable effects in Josephson junctions, it is natural to conjecture that the energy density associated with the underlying primary fluctuations has physical meaning as well: It is a prime candidate for dark energy, being isotropically distributed and temperature independent. There is vacuum energy associated with the measured data in Fig. 1, and it cannot be easily discussed away.

Of course, one has to be careful in the interpretation of the experimental data in Fig. 1. What is really measured with the Josephson junction are real currents produced by real electrons. These currents behave very much in a classical way, however, they are *induced* by zero-point fluctuations. What type of zero-point fluctuations induce the measured currents in the first place is not clear from a theoretical point of view. It could be just ordinary vacuum fluctuations (of QED type), but there is also the possibility of other, new types of vacuum fluctuations, which could potentially underly dark energy and produce a measurable effect in the Josephson junction. What is clear is that if there were no zero-point fluctuations in the first place, then there would also be no linear term in the measured spectrum. Moreover, if the vacuum fluctuations inducing the noise in the junction had a cutoff at some frequency ν_{max}, and would fall quiet above that, the same effect would be observed for the induced physical noise spectrum. In any case, if such a cutoff exists then it would be the *new physics* underlying the cutoff mechanism that makes the system couple to gravity and thus make the corrseponding vacuum energy physically relevant. This is the basic idea underlying the recent suggestion of laboratory tests on dark energy —

checking whether there is a cutoff or not in an extended new version of the Koch experiment [7, 14].

4. Estimating a cosmological cutoff frequency

Already Planck [15] and Nernst [16] formally included zero-point terms in their work. Their formulas yield for the energy density of a collection of oscillators of frequency ν

$$\begin{aligned}\rho(\nu,T) &= \frac{8\pi\nu^2}{c^3}\left[\frac{1}{2}h\nu + \frac{h\nu}{\exp(h\nu/kT)-1}\right] \\ &= \frac{4\pi h\nu^3}{c^3}\coth\left(\frac{h\nu}{2kT}\right). \end{aligned} \tag{4}$$

This just corresponds to the measured spectrum in the Josephson junction experiment, up to a prefactor depending on R. We may split eq. (4) as

$$\rho(\nu,T) = \rho_{vac}(\nu) + \rho_{rad}(\nu,T), \tag{5}$$

where

$$\rho_{vac}(\nu) = \frac{4\pi h\nu^3}{c^3} \tag{6}$$

is induced by zero-point fluctuations, and

$$\rho_{rad}(\nu,T) = \frac{8\pi h\nu^3}{c^3}\frac{1}{\exp(h\nu/kT)-1}. \tag{7}$$

corresponds to the radiation energy density generated by photons of energy $h\nu$. Integration of eq. (6) up to some cutoff ν_c yields

$$\int_0^{\nu_c}\rho_{vac}(\nu)d\nu = \frac{4\pi h}{c^3}\int_0^{\nu_c}\nu^3 d\nu = \frac{\pi h}{c^3}\nu_c^4. \tag{8}$$

This is equivalent to eq. (1) with $m=0$. Integration of eq. (7) over all frequencies yields the well-known Stefan-Boltzmann law

$$\int_0^{\infty}\rho_{rad}(\nu,T)d\nu = \frac{\pi^2 k^4}{15\hbar^3 c^3}T^4. \tag{9}$$

Assuming that vacuum fluctuations (of whatever type) are responsible for dark energy, the necessary cutoff frequency follows from the current astronomical estimates of dark energy density [1, 2]:

$$\rho_{dark} = 0.73\rho_c = (3.9 \pm 0.4) \quad \mathrm{GeV/m^3} \tag{10}$$

Here ρ_c denotes the critical density of a flat universe. From

$$\frac{\pi h}{c^3}\nu_c^4 \simeq \rho_{dark} \tag{11}$$

we obtain

$$\nu_c \simeq (1.69 \pm 0.05) \times 10^{12} \quad \mathrm{Hz}. \tag{12}$$

So if vacuum fluctuations underly dark energy, and if these vacuum fluctuations drive the corresponding quantum oscillators of frequency ν in the resistive element, one would expect to see a cutoff near ν_c in the measured spectrum of the Josephson junction experiment. Because otherwise the corresponding vacuum energy density would exceed the currently measured dark energy density. The frequency ν_c is about 3 times higher than the largest frequency reached in Koch et al.'s experiment of 1982. Future experiments, based on new Josephson junction technology, will be able to reach this higher frequency (see last section).

Suppose a cutoff near ν_c is observed in a future experiment. Then this would represent new physics and the consequences would be far reaching. Frampton [17] has shown that the observation of such a cutoff could even shake some of the assumptions underlying string theory, leading to the possible demise of the so-called string landscape. His argument is based on the fact that if the cutoff is seen in the Josephson experiment then this implies that the dark energy field interacts with the electromagnetic field—which leads to a potential problem for string theory if the physical vacuum, as it is usually assumed, decays by a 1st order phase transition: A small cosmological constant as generated in this context would have decayed to zero by now, contradicting the fact that we do see dark energy right now. In any case, it is interesting that for the first time there seems to be an experiment that can check some of the assumptions underlying the string landscape.

5. New types of vacuum fluctuations?

As said before: If we allow for the possibility of new physics then there are many possible models of dark energy. Let us here consider the possibility that dark energy is produced by new types of vacuum fluctuations with a suitable cutoff. The model was introduced in detail in [18], here we just sketch the main idea.

We start from a homogeneous self-interacting scalar field φ with potential $V(\varphi)$. Most dark energy models are just formulated in a classical setting, but for our approach in terms of vacuum fluctuations we need of course to proceed to a second-quantized theory. We second-quantize our scalar field using the Parisi-Wu approach of stochastic quantization [19], which is a convenient and useful method for our approach. The second-quantized field φ then obeys a stochastic differential equation of the form

$$\frac{\partial}{\partial s}\varphi = \ddot{\varphi} + 3H\dot{\varphi} + V'(\varphi) + L(s,t), \tag{13}$$

where H is the Hubble parameter, t is physical time, s is fictitious time (just a formal coordinate to do quantization) and $L(s,t)$ is Gaussian white noise, δ-correlated both in s and t. The fictitious time s is introduced as a formal tool for 2nd quantization, it has dimensions GeV^{-2}. Quantum mechanical expectations can be calculated as expectations of the above stochastic process for $s \to \infty$. The advantage of the Parisi-Wu approach is that it is very easy and natural to introduce cutoffs in this formulation — by far easier than in the canonical field quantization approach. The simplest way to introduce a cutoff is by making t and s discrete (as in any numerical simulation of a stochastic process). Hence we write

$$s = n\tau \tag{14}$$
$$t = i\delta, \tag{15}$$

where n and i are integers and τ is a fictitious time lattice constant, δ is a physical time lattice constant. Note that the uncertainty relation $\Delta E \Delta t = O(\hbar)$ always implies an effective lattice constant Δt for a given finite energy ΔE. We also introduce a dimensionless field variable Φ_n^i by writing $\varphi_n^i = \Phi_n^i p_{max}$, where p_{max} is some (so far) arbitrary energy scale. The above scalar field dynamics is equivalent to a discrete dynamical system of the form

$$\Phi_{n+1}^i = (1-\alpha)T(\Phi_n^i) + \frac{3}{2}H\delta\alpha(\Phi_n^i - \Phi_n^{i-1}) + \frac{\alpha}{2}(\Phi_n^{i+1} + \Phi_n^{i-1}) + \tau \cdot noise, \tag{16}$$

where the local map T is given by

$$T(\Phi) = \Phi + \frac{\tau}{p_{max}(1-\alpha)} V'(p_{max}\Phi) \tag{17}$$

and α is defined by

$$\alpha := \frac{2\tau}{\delta^2}. \tag{18}$$

For old universes, one can neglect the term proportional to H, obtaining

$$\Phi^i_{n+1} = (1-\alpha)T(\Phi^i_n) + \frac{\alpha}{2}(\Phi^{i+1}_n + \Phi^{i-1}_n) + \tau \cdot noise \tag{19}$$

We now want to construct a field that basically manifests itself as noise: Rather than evolving smoothly it should exhibit strongly fluctuating behavior, so that we may be able to interpret its rapidly fluctuating behaviour in terms of vacuum fluctuations, and possibly in terms of measurable noise in the Josephson junction. As a distinguished example of a φ^4-theory generating such behaviour, let us consider the map

$$\Phi_{n+1} = T_{-3}(\Phi_n) = -4\Phi_n^3 + 3\Phi_n \tag{20}$$

on the interval $\Phi \in [-1, 1]$. T_{-3} is the negative third-order Tchebyscheff map, a standard example of a map exhibiting strongly chaotic behaviour. It is conjugated to a Bernoulli shift, thus generating the strongest possible chaotic behaviour possible for a smooth low-dimensional deterministic dynamical system [20]. The corresponding potential is given by

$$V_{-3}(\varphi) = \frac{1-\alpha}{\tau}\left\{\varphi^2 - \frac{1}{p_{max}^2}\varphi^4\right\} + const, \tag{21}$$

or, in terms of the dimensionless field Φ,

$$V_{-3}(\varphi) = \frac{1-\alpha}{\tau} p_{max}^2 (\Phi^2 - \Phi^4) + const. \tag{22}$$

The important point is that starting from this potential we obtain by second quantization a field φ that rapidly fluctuates on some finite interval, choosing initially $\varphi_0 \in [-p_{max}, p_{max}]$. Since these chaotic fluctuations are bounded, there is a natural cutoff.

The idea is now that the expectation of the potential of this chaotic field (plus possibly kinetic terms) underly the measured dark energy density in the universe. Expectations $\langle\cdots\rangle$ can be easily numerically determined by iterating the dynamics (19) for random initial conditions. One has

$$\langle V_{-3}(\varphi)\rangle = \frac{1-\alpha}{\tau} p_{max}^2 (\langle\Phi^2\rangle - \langle\Phi^4\rangle) + const, \tag{23}$$

which for $\alpha = 0$ can be analytically evaluated [20] to give

$$\langle V_{-3}(\varphi)\rangle = \frac{1}{8}\frac{p_{max}^2}{\tau} + const. \tag{24}$$

To reproduce the currently measured dark energy, we only need to fix the ratio of the parameters τ and p_{max} as

$$\frac{p_{max}^2}{\tau} \sim \rho_\Lambda \sim m_\varphi^4 \tag{25}$$

This is the simplest model of noise-like vacuum fluctuations with a suitable finite cutoff one can think of, a chaotic scalar field theory underlying the cosmological constant. It is easy to show [18] that for $\alpha = 0$ the equation of state of this field is $w = -1$. For small α, it is close to $w = -1$. In this model, there is no reason why the deterministic chaotic noise represented by this rapidly fluctuating field should not be able to influence electric charges. Hence these chaotic fluctuations may well induce a measurable noise spectrum in Josephson junctions.

6. A new quantum noise experiment

In [7] we suggested to repeat the experiment of Koch et al. with new types of Josephson junctions that are capable of reaching higher frequencies. This new experiment will now be built, the grant has just been allocated [21]. Warburton, Barber, and Blamire are planning two different versions of the experiment. One is based on nitride junctions, the other one on cuprate junctions. The maximum frequency that can be reached is determined by the gap energy of the Josephson junction under consideration, and the above materials provide the possibility to reach the cosmologically interesting frequency of $\nu_c \approx 1.7$ THz and even exceed it. The new technology suggested by Warburton et al. is more sophisticated than that of conventional Niobium-based Josephson junctions. With conventional Niobium-based junctions one would probably be only able to reach 1.5 THz in the measurements since their gap energy is too small. By performing experiments on both the nitrides and the cuprates there will be two independent high frequency measurements of the quantum noise spectrum in two very different material systems. So in about 3 years time we should know whether there is any unusual behaviour of zero-point fluctuations near ν_c, which could possibly be related to dark energy. Whatever the outcome of this new experiment, the result will be interesting: If a cutoff is observed it will revolutionize our understanding of dark energy. If a cutoff is not observed, it will show that the vacuum fluctuations measured with the Josephson junction are definitely not gravitationally active.

[1] Bennett, C.L. et al. (2003). Astrophys. J. Supp. Series 148, 1 (astro-ph/0302207)
[2] Spergel, D.N. et al. (2003). Astrophys. J. Supp. Series 148, 148 (astro-ph/0302209)
[3] Peebles, P.J.E. and Ratra, B. (2003). Rev. Mod. Phys. 75, 559 (astro-ph/0207347)
[4] Weinberg, S. (2000). astro-ph/0005265
[5] Trodden, M. and Caroll S. (2004). astro-ph/0401547
[6] Padmanabhan, T. (2003). Phys. Rep. 380, 235 (hep-th/0212290)
[7] Beck, C. and Mackey, M.C., Phys. Lett. B 605, 295 (2005) (astro-ph/0406504)
[8] Gardiner, C.W. (1991). *Quantum Noise*, Springer, Berlin
[9] Koch, R.H., van Harlingen, D. and Clarke J. (1980). Phys. Rev. Lett. 45, 2132
[10] Koch, R.H., van Harlingen, D. and Clarke J. (1982). Phys. Rev. B 26, 74
[11] Levinson, Y. (2003). Phys. Rev. B 67, 184504
[12] Jetzer, P. and Straumann, N., Phys. Lett. B 606, 77 (2005) (astro-ph/0411034)
[13] Bresst, G. et al., Phys. Rev. Lett. 88, 041804 (2002)
[14] Ball, P., Nature 430, 126 (2004)
[15] Planck, M. (1914). *The Theory of Heat Radiation*. P. Blakiston's Son & Co.
[16] Nernst, W. (1916). Verh. Dtsch. Phys. Ges., 18, 83
[17] Frampton, P.H., hep-th/0508082
[18] Beck, C. (2004) Phys. Rev. D 69, 123515 (astro-ph/0310479)
[19] Parisi, G. and Wu Y.S. Sci. Sin 24, 483 (1981)
[20] Beck C., *Spatio-temporal Chaos and Vacuum Fluctuations of Quantized Fields*, World Scientific, Singapore (2002)
[21] Warburton, P.A., Barber, Z., and Blamire, M., *Externally-shunted high-gap Josephson junctions: Design, fabrication and noise measurements*. EPSRC grants EP/D029783/1 and EP/D029872/1

Following is the title page of one of Koch's articles (1st of 14 pages) which details his measurement of zero-point fluctuations. Koch measured current noise (Fig. 6 has been reproduced by Dr. Christian Beck) in the high gigahertz range (10^{11} Hz) and current in the 10^{-10} ampere range, which is a tenth of one nanoampere, twenty five years ago. - TV

PHYSICAL REVIEW B VOLUME 26, NUMBER 1 1 JULY 1982

Measurements of quantum noise in resistively shunted Josephson junctions

Roger H. Koch, D. J. Van Harlingen,* and John Clarke
Department of Physics, University of California, Berkeley, California 94720
and Materials and Molecular Research Division, Lawrence Berkeley Laboratory,
Berkeley, California 94720
(Received 19 November 1981)

Measurements have been made of the low-frequency spectral density of the voltage noise in current-biased resistively shunted Josephson tunnel junctions under conditions in which the noise mixed-down from frequencies near the Josephson frequency (ν_J) to the measurement frequency ($\ll \nu_J$) is in the regime $h\nu_J > k_B T$. In this limit, quantum corrections to the mixed-down noise are important. The spectral densities measured on junctions with current-voltage characteristics close to the Stewart-McCumber model were in excellent agreement with the predicted values, with no fitted parameters. The mixed-down noise for a wide range of bias voltages was used to infer the spectral density of the current noise in the shunt resistor at frequency ν. With no fitted parameters, this spectral density at frequencies up to 500 GHz was in excellent agreement with the prediction $(2h\nu/R)\coth(h\nu/2k_B T)$. The presence of the zero-point term, $2h\nu/R$, at frequencies $h\nu > k_B T$ was clearly demonstrated. The current-voltage characteristics of a junction with $\beta_L = 2\pi L_s I_0/\Phi_0 \sim 1$ and $\beta_c = 2\pi I_0 R^2 C/\Phi_0 \ll 1$, where I_0 is the critical current, C is the junction capacitance, and L_s is the shunt inductance, showed structure at voltages where the Josephson frequency was near a subharmonic of the $L_s C$ resonant frequency. The additional nonlinearity of the I-V characteristic caused mixing down of noise near higher harmonics of the Josephson frequency, thereby greatly enhancing the voltage noise. The measured noise was in good agreement with that predicted by computer simulations.

I. INTRODUCTION

The effects of thermal noise on a resistively shunted[1,2] Josephson[3] junction (RSJ) have been extensively studied. The theories assume that the noise originates as Nyquist noise in the shunt resistor R. The junction is modeled as a particle moving in a tilted periodic potential, and the effect of the noise current is to induce random fluctuations in the angle of tilt. These fluctuations have two effects. First, they enable the phase of the junction to slip by 2π when the bias current I is less than the noise-free critical current I_0, thereby producing a voltage pulse across the junction. This effect produces noise rounding of the I-V characteristics at low voltages, V; the noise rounding has been calculated by Ambegoakar and Halperin[4] and Vystavkin *et al.*[5] for the case $C = 0$ (C is the capacitance of the junction). Subsequently, Kurkijärvi and Ambegoakar[6] and Voss[7] computed the case $C \neq 0$. Second, the fluctuations generate a voltage noise when the junction is current biased at a nonzero voltage. Likharev and Semenov[8] and Vystavkin *et al.*[5] showed that for the $C = 0$ case in the limit $h\nu_J \ll k_B T$ ($\nu_J = 2eV/h$ is the Josephson frequency) and for frequencies much less than ν_J, the spectral density of the voltage noise is given by

$$S_v(0) = \frac{4k_B T R_D^2}{R}\left[1 + \frac{1}{2}\left(\frac{I_0}{I}\right)^2\right] . \qquad (1.1)$$

Here, R_D is the dynamic resistance. This result was derived on the assumption that the noise is sufficiently small that one can neglect departures of the I-V characteristic from that of the ideal RSJ,[1,2]

$$V = R(I^2 - I_0^2)^{1/2} . \qquad (1.2)$$

Thus, Eq. (1.1) is not valid in the noise-rounded region $I \lesssim I_0$. Voss[7] and Koch and Clarke[9] computed the noise for the case $C \neq 0$. Experimental results are in good agreement with calculations for both the noise rounding[10] and voltage noise.[11]

For a junction voltage-biased on a self-resonant step, Stephen[12] has calculated the contribution of pair current fluctuations to the linewidth of the Josephson radiation. This noise arises from photon number fluctuations (including zero-point fluctuations) in the lossy cavity formed by the junction, and is not intrinsic to the tunneling of Cooper pairs in a nonresonant junction. Experimental re-

The last sentence on this page continues the introductory summary: "Experimental results are in good agreement with the predictions." You can purchase the entire article from American Physics Society online at: http://prola.aps.org/abstract/PRB/v26/i1/p74_1

The Alcubierre 'Warp Drive'

http://www.astro.cf.ac.uk/pub/Miguel.Alcubierre/

http://www.sr.bham.ac.uk/~act/FTL/

The theoretical physicist Miguel Alcubierre was born in Mexico City, where he lived until 1990 when he travelled to Cardiff in the UK to enter graduate school at the University of Wales. He received his PhD from that institution in 1993 for research in numerical general relativity, solving Einstein's gravitational equations with fast computers. In 1994 Alcubierre produced an idea which grew from his work in general relativity. His paper describes an unusual solution to Einstein's equations of general relativity, described in the title as a 'warp drive', and in the abstract as 'a modification of space time in a way that allows a space ship to travel at an arbitrarily large speed'.

In the context of special relativity, the speed of light is the absolute speed limit of the universe for any object having a real mass, for two reasons. First, giving a fast object even more kinetic energy has the main effect of causing an increase in mass-energy rather than speed, with mass-energy approaching infinite as speed tends to the velocity of light. By this mechanism, relativistic mass increase limits massive objects to sub-light velocities.

There is also a second faster than light (FTL) prohibition. Special relativity is based on the treatment of all reference frames (i.e., co-ordinate system moving at some constant velocity) with perfect even-handedness and 'democracy'. Therefore, FTL communication is implicitly ruled out by special relativity because it could be used to perform 'simultaneity tests' of the readings of separated clocks which would reveal the preferred or 'true' reference frame of the universe. The existence of such a preferred frame is in conflict with special relativity.

General relativity treats special relativity as a restricted sub-theory that applies locally to any region of space sufficiently small that its curvature can be neglected. General relativity does not forbid faster-than-light travel or communication, but it does require that the local restrictions of special relativity must apply. In other words, light speed is the local speed limit, but the broader considerations of general relativity may provide a way of circumventing this local statute. One example of this is a wormhole connecting two widely separated locations in space.

Another example of FTL in general relativity is the expansion of the universe itself. As the universe expands, new space is being created between any two separated objects. The objects may be at rest with respect to their local environment and with respect to the cosmic microwave background, but the distance between them may grow at a rate greater than the velocity of light. According to the standard model of cosmology, parts of the universe are receding from us at FTL speeds, and therefore are completely isolated from us. As the rate of expansion of the universe diminishes due to the pull of gravity, remote parts of the universe that have been out of light-speed contact with us since the Big Bang are coming over the lightspeed horizon and becoming newly visible to our region of the universe.

Alcubierre has proposed a way of overcoming the FTL speed limit that is analogous to the expansion of the universe, but on a more local scale. He has developed a metric that describes a region of flat space surrounded by a 'warp' that propels it forward at any arbitrary velocity, including FTL speeds. Alcubierre's warp is constructed of hyperbolic tangent functions which create a very specific distortion of space at the edges of the flat-space volume. In effect, new space is rapidly being created (like an expanding universe) at the back side of the moving volume, and existing space is being annihilated (like a universe collapsing to a Big Crunch) at the front side of the moving volume. Thus, a space ship within the volume of the

Alcubierre warp (and the volume itself) would be pushed forward by the expansion of space at its rear and the contraction of space in front.

The Alcubierre warp metric has some unusual aspects. Since a ship at the centre of the moving volume of the metric is at rest with respect to locally flat space, there are no relativistic mass increase or time dilation effects. The on-board spaceship clock runs at the same speed as the clock of an external observer, and that observer will detect no increase in the mass of the moving ship, even when it travels at FTL speeds. Moreover, Alcubierre has shown that even when the ship is accelerating, it travels on a free-fall geodesic. In other words, a ship using the warp to accelerate and decelerate is always in free fall, and the crew would experience no accelerational gee-forces. Enormous tidal forces would be present near the edges of the flat-space volume because of the large space curvature there, but by suitable specification of the metric, these would be made very small within the volume occupied by the ship.

There are two 'catches' in the Alcubierre warp drive scheme. The first is that, while his warp metric is a valid solution of Einstein's equations of general relativity, we have no idea how to produce such a distortion of space-time. Its implementation would require the imposition of radical curvature on extended regions of space. Within our present state of knowledge, the only way of producing curved space is by using mass, and the masses we have available for works of engineering lead to negligible space curvature. Moreover, even if we could do engineering with mini black it is not clear how an Alcubierre warp could be produced.

Alcubierre has also pointed out a more fundamental problem with his warp drive. General relativity provides a procedure for determining how much energy is implicit in a given metric (or curvature of space-time). He shows that the energy density is negative, large, and proportional to the square of the velocity with which the warp moves forward. This means that the weak, strong, and dominant energy conditions of general relativity are violated, which can be taken as arguments against the possibility of creating a working Alcubierre drive. Alcubierre, following the lead of wormhole theorists, argues that quantum field theory permits the existence of regions of negative energy density under special circumstances, and cites the Casimir effect as an example. Thus, the situation for the Alcubierre drive is similar to that of stable wormholes: they are solutions to the equations of general relativity, but one would need "exotic matter" with negative mass-energy to actually produce them, and we have none at the moment.

Ref.: Alcubierre, Miguel. "The warp drive: hyper-fast travel within general relativity" 1994 Journal Classical and Quantum Gravity. 11 L73-L77

Derivation of ZPE, the Lowest Energy State

THOMAS VALONE
Integrity Research Institute

Look how simple this fundamental ZPE argument is:

Interestingly, the ground state energy of a simple harmonic oscillator (SHO) model can also be used to find the average value for zero-point energy. *This is a valuable exercise to show the fundamental basis for zero-point energy parton oscillators.* The harmonic oscillator is used as the model for a particle with mass m in a central field. The **uncertainty principle** provides the only requisite for a derivation of the minimum energy of the simple harmonic oscillator, utilizing the equation for kinetic and potential energy,

$$E = p^2/2m + \tfrac{1}{2}\, m\, \omega^2\, x^2 \,. \qquad \text{Eq. 1}$$

Solving the uncertainty relation

$$\Delta x\, \Delta p \geq h/4\pi \qquad \text{Eq. 2}$$

for p, one can substitute it into Equation (1). Using a calculus approach, one can take the derivative with respect to x and set the result equal to zero. A solution emerges for the value of x that is at the minimum energy E for the SHO. This x value can then be placed into the minimum energy SHO equation where the potential energy is set equal to the kinetic energy.

The ZPE solution yields **½hf** for the minimum energy E, which is the desired answer in Planck's Second Law.[263]

Is Permanent Magnetism Connected to Zero Point Energy?

Thomas F. Valone
Integrity Research Institute

Introduction

People often ask me if permanent magnets can be a source of energy for the future. My answer is that the "magnetic gradient" (d**B**/dx) is an untapped potential energy source that is just like all of the other gradients we depend upon for energy, such as the thermal gradient in heat pumps, the voltage gradient in our wall sockets, the gravity gradient in our hydroelectric plants, the pressure gradient in our water pipes, etc. We just have not designed and commercially perfected magnetic gradient machines that work continuously. However, there are several promising patents that offer a template for what such a motor or generator would look like, such as the Hartman magnetic gradient track detailed in the expired 1980 US patent #4,215,330 by Emil T. Hartman. The Hartman invention is designed to propel a steel ball upward along a 10-degree inclined plane by the magnetic forces of permanent (cylindrical) bar magnets and then, without means other than gravity, cause the ball to drop from the end of the plane entirely out of the magnetic field. Another similar invention is the set of Kuroda Takeshi Japanese patents (from Kure Tekkosho Co., e.g. JP55144783, JP55114172, JP55061273, etc.) also portrayed in the *Popular Science* article of June 1979. Lastly, one of the more complex magnetic designs is the Roschin and Godin patent #6,822,361 (assigned to Energy & Propulsion Systems LLC), which I also helped draft, is multi-rotor homopolar motor design. For more information on this last example, the 2001 Joint Propulsion Conference paper, "An Experimental Investigation of the Physical Effects in a Dynamic Magnetic

System," by the inventors is a good summary of the experimental findings (AIAA 2001-3660).

Physics of Magnetism

If any of the above-mentioned patents become commercially viable in the near future, the question will arise, "Where is the energy coming from?" My answer is that the magnetic field has energy and can be regarded as an energy source in physics, just as the electric field is depended upon for the same purpose. However, that standard hand-waving, textbook-thumping argument is not very satisfying. Instead, I needed to look further to the activity of the electron in the atoms that are creating the magnetic field in the first place. The book by Soshin Chikazumi entitled, *Physics of Magnetism* (J. Wiley, 1964) is a classic text that has stood the test of time. Here we find the quantum mechanics of electron spin **S** and electron orbital **L** magnetic moments. In magnetic atoms, the author points out there is a "magnetic anisotropy" that is created between the sum of the spin and orbital magnetic moments, which varies from one atom to another. However, as a general rule, "the **origin of ferromagnetism** is not an orbital motion but *a spin motion of the electron*" says the author (p.48). In fact, just spinning a rod of Permalloy around its axis at sufficient speed will magnetize it to saturation, thus demonstrating a macroscopic effect of electron spin and magnetism.

The total angular momentum **J**, which is closely related to the total magnetic moment of a free atom, is the resultant or vector sum of the coupling between the spin and orbital magnetic moments, as you would suspect, **J = L + S**.

So far so good. However, we would like to find out if there is a physics principle that links **J** to the zero point energy (ZPE) of the vacuum. Chikazumi points out (P. 57) that "when the electron is traveling around the nucleus, the electron sees the nucleus as if the nucleus is traveling around the electron itself and feels the magnetic field caused by the circulating nuclear charge." This will be a positive interaction (reinforcing) between the electron and the nucleus if the electron shell is less than half filled. Of course, it is noted that only the protons in the nucleus, which are positively charged, contribute to the nuclear charge.

Where Does The Proton Really Get Its Spin?

Now, since **J** depends so much on the proton, this leads us to an ancillary question that could be asked, "Where does the proton really get its spin?" Luckily, this is the title of an article published in Physics Today (Sept. 1995, p. 24) which states that "quark spins appear to account for 20-30% of the spin of the proton or neutron." (There are three quarks that make up every proton.) "But the fact remains that much of the nucleon's spin lies elsewhere." This leads the author, Robert Jaffe, professor of physics at MIT's Center for Theoretical Physics, to conclude that the "sea of quark-antiquark pairs" is a strong candidate, from the Dirac sea of virtual particle pairs. His finding is reminiscent of Beck, Koch and Davis who argue based on experiment, that ZPE manifests within solid state devices. Jaffe states (p. 29) that "Because strange quarks in the nucleon would appear only in pairs, the nonvanishing of Δs is *prima facie* evidence that virtual quark-antiquark pairs carry a significant fraction of its spin."

So far we have the proton spinning (figuratively speaking) around the electron which affects the total magnetic moment. But *does the proton spinning around itself contribute to the electron spin*? That answer comes from a standard quantum mechanics textbook, *Lectures on Quantum Mechanics* by Gordon Baym (Benjamin/Cummings Pub. 1969). To gain a perspective, Baym points out that the mass ratio m/M of the electron vs. the nucleus is about 1/10,000 or smaller. "As a consequence the zero-point motion of the nuclei is far less than that of the electrons" (p.469). As he compares the vibrational energy states of the electron vs. the nucleus (both are quantized of course), he notes that the nuclear vibration energies, using E=hf, are a factor of $m/(M)^{1/2}$ smaller than the electron vibration energies. However, as Baym compares the zero-point momentum of a nucleus, using a standard harmonic oscillator well model, he finds "this is about ten times greater than electronic momenta" (p.470). This demonstrates how much more ZPE of the vacuum is being exchanged with the nucleus, since kinetic energy is simply momentum divided by twice the mass (P/2M).

All of this nuclear physics shows the relative surprises that the nucleus and the electron hold in store in a molecule or atom, often exchanging energy between them. Using the Schroedinger equation, Baym notes that the nuclear wave function is a product of translational, rotational, and vibrational parts and that "we should note that the nuclear modes and frequencies depend on the particular electronic state" (p.474). In other words, the state of the electrons, including its rotational spin state, affects the nucleus and vice versa. Each system of electrons or nucleus is a potential function for the other one's state.

Peter Milonni, author of *The Quantum Vacuum* (Academic Press, 1994) goes even further. After calculating the rate at which the atom absorbs energy from the vacuum field (the all-pervading field of zero point energy), he also calculates the rate at which the atom loses energy due to radiation. Equating both of them for equilibrium purposes, he finds that the equation simplifies to the Bohr quantization condition for the ground state of the hydrogen atom! Therefore, one is forced to conclude, "We now know that the vacuum field is in fact formally necessary for the stability of atoms in quantum theory" (p.81).

Conclusion

This paper serves only to introduce the argument that the spin and orbital motion of the electron is intimately coupled to the nuclear energy states, which also include rotational components. Not only have we shown that the proton spin derives from the vacuum field of quark-antiquark pairs, but that the atom itself is deriving energy from the vacuum ZPE field.

Furthermore, it has been also shown that it is a well-known fact in quantum mechanics that the electron and the nucleus exchange energy and depend upon each other in an intimate fashion, acting as a coupled system of oscillators. Therefore, as we ask if there is a macroscopic manifestation of all of this ZPE exchange, especially with regard to the **J** of the electron, it can be easily concluded that there is a clear pathway for ZPE to be the sustaining energy source for all energy states of the atom, including the angular momentum **J** of the electron. With that physics principle established, the macroscopic magnetic field of a permanent magnet, which is totally attributed to the **J** of the electron, can be said to be sustained by the vacuum ZPE field. Therefore, as entirely permanent magnet motors, generators, and actuators become commercialized, it will no longer be a mystery as to where the energy is coming from. Furthermore, these magnet-powered devices cannot be mistaken for *perpetuum mobiles*.

If that was not enough proof, a Casimir expert, Dr. Iannuzzi, from Vrie University, Amsterdam, has been teaching "quantum electrodynamical torque" as an interesting phenomenon that has not received adequate attention, in reference to an experiment running at Harvard University. This quantum torque, which is a twisting force, depends directly on ZPE and is summarized in the figure below. Davide Iannuzzi's 2006 web posting of this slide show can be found at http://www.inrim.it/events/docs/Casimir/Iannuzzi.pdf

Quantum electrodynamical torque

Birefringent materials: reflection, absorption, transmission depend on orientation

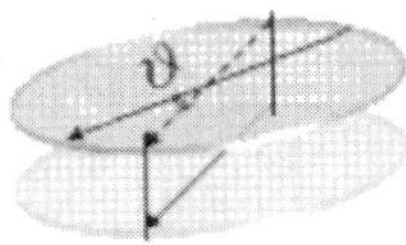

Zero-point energy depends on the orientation too!

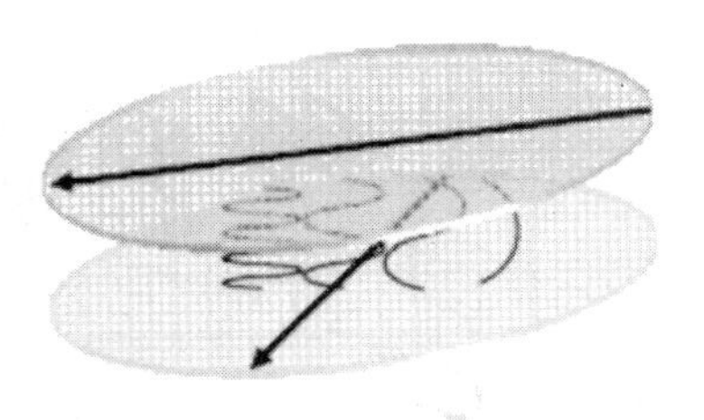

$$M = -\frac{\partial E}{\partial \vartheta}$$

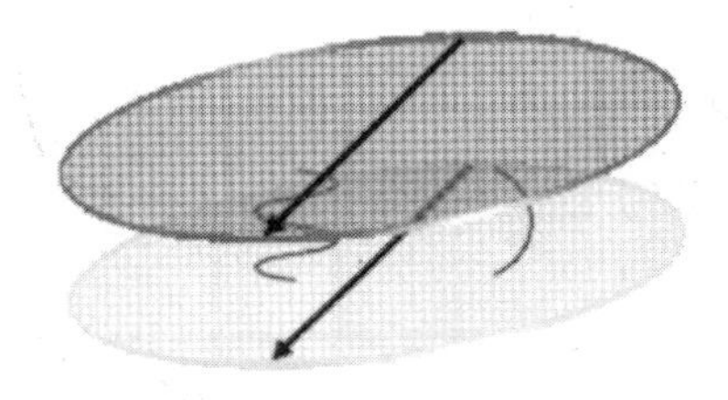

V. A. Parsegian and G. H. Weiss, *J. Adhesion* **3** (1972) 259 (non-retarded limit)
Y. Barash, *Izvestiya vuzov, Radiofizika*, Tom **XXI** (1978) 1637

Quantum Electrodynamical torque

davide iannuzzi - vrije universiteit amsterdam

Proposed Use of Zero Bias Diode Arrays

Proceedings of Space, Propulsion and Energy Sciences International Forum (SPESIF), Workshop on Future Energy Sources. American Institute of Physics, American Institute of Aeronautics and Astronautics, Huntsville AL, February 24, 2009

Proposed Use of Zero Bias Diode Arrays as Thermal Electric Noise Rectifiers and Non-Thermal Energy Harvesters

Thomas F. Valone

Integrity Research Institute
5020 Sunnyside Avenue, Suite 209
Beltsville MD 20705
301-220-0440; IRI@starpower.net

Abstract. The well known built-in voltage potential for some select semiconductor p-n junctions and various rectifying devices is proposed to be favorable for generating DC electricity at "zero bias" (with no DC bias voltage applied) in the presence of Johnson noise or 1/f noise which originates from the quantum vacuum (Koch, 1982). The 1982 Koch discovery that certain solid state devices exhibit measurable quantum noise has also recently been labeled a finding of dark energy in the lab (Beck, 2004). Tunnel diodes are a class of rectifiers that are qualified and some have been credited with conducting only because of quantum fluctuations. Microwave diodes are also good choices since many are designed for zero bias operation. A completely passive, unamplified zero bias diode converter/detector for millimeter (GHz) waves was developed by HRL Labs in 2006 under a DARPA contract, utilizing a Sb-based "backward tunnel diode" (BTD). It is reported to be a "true zero-bias diode". It was developed for a "field radiometer" to "collect thermally radiated power" (in other words, 'night vision'). The diode array mounting allows a feed from horn antenna, which functions as a passive concentrating amplifier. An important clue is the "noise equivalent power" of 1.1 pW per root hertz and the "noise equivalent temperature difference" of 10°K, which indicate sensitivity to Johnson noise (Lynch, et al., 2006). There also have been other inventions such as "single electron transistors" that also have "the highest signal to noise ratio" near zero bias. Furthermore, "ultrasensitive" devices that convert radio frequencies have been invented that operate at outer space temperatures (3 degrees above zero point: 3°K). These devices are tiny nanotech devices which are suitable for assembly in parallel circuits (such as a 2-D array) to possibly produce zero point energy direct current electricity with significant power density (Brenning et al., 2006). Photovoltaic p-n junction cells are also considered for possible higher frequency ZPE transduction. Diode arrays of self-assembled molecular rectifiers or preferably, nano-sized cylindrical diodes are shown to reasonably provide for rectification of electron fluctuations from thermal and non-thermal ZPE sources to create an alternative energy DC electrical generator in the picowatt per diode range.

Keywords: Diodes, Rectifiers, Energy Harvesting, Quantum Vacuum, Zero Point Energy, Direct Current, Nonthermal Noise, 1/F Noise, Shot Noise, Johnson Noise
PACS: 07.50.Hp, 05.40.-a, 03.75.Lm, 85.35.Gv, 81.07.N

INTRODUCTION

The US currently spends between 5 and 10 cents per kilowatt-hour (kWh) depending upon whether we are a residential or commercial customer. Furthermore, the US Electric Power Industry generates approximately 4,000 billion kWh on an annual basis (www.eia.doe.gov). These figures indicate that electricity consumption is about a $300 billion market commanded by the public utilities. It is proposed that distributed single cubic-meter electricity generating units may become a reality in the near future with the emergence of zero point energy (ZPE) rectifiers deployed in the form of three-dimensional arrays. This event is predicted to create a disruptive effect on the public utilities, while it empowers ordinary individuals from all walks of life including third world countries, opening up vast areas of the world that are presently uninhabitable due to the lack of on-site energy generation capability.

This paper of mine is published by the American Institute of Physics. More information about the conference is available at http://www.ias-spes.org/SPESIF.html - TV

Glossary

Following are terms that are used throughout the book:

1. Bremsstrahlung: Radiation caused by the deceleration of an electron. Its energy is converted into light. For heavier particles the retardations are never so great as to make the radiation important.[264]
2. Dirac Sea: The physical vacuum in which particles are trapped in negative energy states until enough energy is present locally to release them.
3. Energy: The capacity for doing work. Equal to power exerted over time (e.g. kilowatt-hours) or force times distance (i.e. Newton-meters). It can exist in linear or rotational form and is quantized in the ultimate part. It may be conserved or not conserved, depending upon the system considered. Mostly all terrestrial manifestations can be traced to solar origin, except for zero-point energy.
4. Lamb Shift: A shift (increase) in the energy levels of an atom, regarded as a Stark effect, due to the presence of the zero-point field. Its explanation marked the beginning of modern quantum electrodynamics.
5. Parton: The fundamental theoretical limit of particle size thought to exist in the vacuum, related to the Planck length (10^{-35} meter) and the Planck mass (22 micrograms), where quantum effects dominate spacetime. Much smaller than subatomic particles, it is sometimes referred to as the charged point particles within the vacuum that participate in the ZPE Zitterbewegung.
6. Planck's Constant: The fundamental basis of quantum mechanics which provides the measure of a quantum ($h = 6.6 \times 10^{-34}$ joule-second), it is also the ratio of the energy to the frequency of a photon.
7. Quantum Electrodynamics: The leading quantum theory of light as electromagnetic radiation, in wave and particle form, as it interacts with matter. Abbreviated "QED."
8. Quantum Vacuum: A characterization of empty space by which physical particles are unmanifested or stored in negative energy states. Also called the "physical vacuum."

9. Uncertainty Principle: The rule or law that limits the precision of a pair of physical measurements in complimentary fashion, e.g. the position and momentum, or the energy and time, forming the basis for zero-point energy.
10. Virtual Particles: Physically real particles emerging from the quantum vacuum for a short time determined by the uncertainty principle. This can be a photon or other particle (most likely an electron) in an intermediate state which, in quantum mechanics (Heisenberg notation) appears in matrix elements connecting initial and final states. Energy is not conserved in the transition to or from the intermediate state. Also known as a virtual quantum.
11. Zero-point energy: The non-thermal, ubiquitous kinetic energy (averaging ½hf or 'half-photon') that is manifested even at zero degrees Kelvin, abbreviated as "ZPE." Also called vacuum fluctuations, zero-point vibration, residual energy, quantum oscillations, the vacuum electromagnetic field, virtual particle flux, and recently, dark energy.
12. Zitterbewegung: An oscillatory motion of an electron, exhibited mainly when it penetrates a voltage potential, with frequency greater than 10^{21} Hertz. It can be associated with pair production (electron-positron) when the energy of the potential exceeds $2mc^2$ (m = electron mass). Also generalized to represent the rapid oscillations associated with zero-point energy.

ZPE and Casimir Links on Web

"Design Manual for Micromachines Using Casimir Forces" –Jordan Maclay http://www.quantumfields.com/staif-2000paper.PDF

Dr. Fabrizio Pinto's company website – http://www.interstellartechcorp.com

The Casimir Effect - undergraduate MIT physics paper - http://web.mit.edu/ttorres/Public/The%20Casimir%20Effect.pdf

Casimir Effect slide show - experiments shown UC Berkeley - http://socrates.berkeley.edu/~budker/Physics250_Spring0607/Corsini_Casimir.ppt

"Precision Casimir Force Measurement" by Mohideen and Roy, U of Calif, Riverside - http://arxiv.org/pdf/physics/9805038

Casimir Biography - http://www.iop.org/EJ/article/0143-0807/22/4/320/ej1420.pdf

"Casimir Effect and Vacuum Fluctuations" - Ohio University paper - http://plato.phy.ohiou.edu/~ulloa/611-612/612papers/Trang%20Nguyen--Casimir%20Effect.pdf

"Measurement of the Casimir Force between Parallel Metallic Surfaces" - Dartmouth College - http://www.dartmouth.edu/~ongroup/publications/PRL_88_041804.pdf

"Vacuum Fluctuations and the Casimir Force Theory" - Swarthmore College - http://www.sccs.swarthmore.edu/users/02/lisal/physics/presentations/casimir.pdf

"The Classical Vacuum" [Zero-Point Energy] - reprint of Boyer article from Scientific American, 1985 http://www.padrak.com/ine/ZPESCIAM2.html

Zero Point Energy Insights - idealized report http://www.secret-solutions.com/zpe.htm

"Resource Letter on Casimir Force" by Dr. Lamoreaux, Los Alamos Nat. Lab http://link.aip.org/link/?AJPIAS/67/850/1

Graphical Low Tech Discussion of Thrust and Electrical Power from Rectifying ZPE http://www.keelynet.com/zpe/chaos.htm

Casimir Force Experiments slides - Iannuzzi - Vrije University Netherlands - 2006 (amazing and informative) http://www.inrim.it/events/docs/Casimir/Iannuzzi.pdf

Casimir Energies for Single Cavities - Turkey J of Physics http://journals.tubitak.gov.tr/physics/issues/fiz-06-30-4/fiz-30-4-15-0607-21.pdf

"Nothing Like a Vacuum" - London Sunday Telegraph popular article 1995 http://www.calphysics.org/haisch/matthews.html

Some Emerging Possibilities - NASA Warp Drive When? website http://www.nasa.gov/centers/glenn/research/warp/possible.html

"Zero Point Fields, Gravitation and the New Physics" - U of Waterloo http://www.calphysics.org/articles/wesson.pdf

Casimir Interactions slides - Wirzba - Germany http://www.itkp.uni-bonn.de/~wirzba/ps/casimir2.pdf

Drs. Hal Puthoff and Eric Davis, Inst. For Advanced Studies at Austin, http://www.earthtech.org

ZPEnergy website devoted to ZPE and free energy – http://www.zpenergy.com

Tom Valone's Integrity Research Institute ZPE webpage with DVDs and publications – http://users.erols.com/iri/ZPENERGY.html

Tom Valone's complete lecture on the details and practical uses of zero point energy at a recent Extraordinary Technology conference: http://video.google.com/videoplay?docid=-5738531568036565057

Index

References

[1] Kuhn, Thomas. *The Structure of Scientific Revolutions* University of Chicago Press, 1970

[2] Iannuzzi, Davide. Vrije University, Amsterdam, Netherlands, 2006 http://www.inrim.it/events/docs/Casimir/Iannuzzi.pdf

[3] Grace, Tom. *Quantum*, Warner Books, 2000

[4] Miller, William and Langdon Morris. *Fourth Generation R & D: Managing Knowledge, Technology and Innovation*, John Wiley & Sons, Inc. 1999

[5] Yam, Philip. "Exploiting Zero-Point Energy." Scientific American. December, 1997, p. 82

[6] Barrow, John. *The Book of Nothing.* Pantheon Books, New York, 2000, p. 210

[7] Silvertooth, E.W. et al., "A New Michelson-Morley Experiment" *Physics Essays*, V. 5, 1992, p. 82-89

[8] Lapedes, Daniel, Editor, *McGraw-Hill Dictionary of Physics and Mathematics*, McGraw-Hill, 1978

[9] Walt Disney Productions

[10] Wolf, Fred Alan. *Star Wave: Mind, Consciousness and Quantum Physics*, Collier Books, 1984, p. 120

[11] Forward, Robert. "Extracting Electrical Energy from the Vacuum by Cohesion of Charged Foliated Conductors," *Physical Review B*, Vol. 30, No. 4, 1984, 30,1700

[12] Bortman, Henry. "Energy Unlimited" *New Scientist*, V. 165, Issue 2222, p. 32

[13] Jaekel, Marc-Thierry et al. "Movement and fluctuations of the vacuum" *Rep. Prog. Phys.*, V. 60, 1997, p. 867

[14] ibid., p. 879-880

[15] Bordag, M. et al., "New Developments in the Casimir Effect" *Physics Reports*, V. 353, 2001, p. 4 (The authors refer to the "infinite vacuum

energy" and "infinite zero-point energy" on page 4 of this 205-page report on the Casimir force.)

[16] Milonni, Peter. *The Quantum Vacuum*, Academic Press, San Diego, 1994, p. 56

[17] Milonni, p. 53

[18] Using the speed of light (c) as a constant, the equation relating vibration frequency (f)and wavelength (λ) is $c=f\lambda$. This makes wavelength and frequency <u>inversely</u> proportional. This equation can be used, for example, to find the fundamental resonant frequency of the earth-ionosphere cavity, using $\lambda=25{,}000$ miles as the circumference of the earth (for one wavelength) and 186,000 miles per second for c. (Answer: f = 7 ½ Hz)

[19] "Vacuum fluctuations remain a matter of debate, mainly because their energy is infinite. More strikingly, their energy per volume is infinite." Marc-Thierry Jaekel et al. "Movement and Fluctuations in the Vacuum" Rep. Prog. Phys. V. 60, 1997, p. 863
"I must say that I am very dissatisfied with the situation, because this so-called 'good theory' does involve neglecting infinities which appear in its equations, neglecting them in an arbitrary way. This is just not sensible mathematics. Sensible mathematics involves neglecting a quantity when it turns out to be small—not neglecting it just because it Is infinitely great and you do not want it!" Paul M. Dirac, Directions in Physics, edited by H. Hora et al., J. Wiley & Sons, 1978, p. 36

[20] Yam, Philip. "Exploiting Zero-Point Energy." Scientific American. December, 1997, p. 82

[21] Forward, Robert. "An Introductory Tutorial on the Quantum Mechanical Zero Temperature Electromagnetic Fluctuations of the Vacuum," *Phillips Laboratory Report #PL-TR 96-3004*, 1996

[22] Vacca, John, *The World's 20 Greatest Unsolved Problems*, Prentice Hall, 2005.

[23] "Free Energy: Race to Zero Point" 2 hour documentary DVD and VHS video, for which this author served as "technical consultant." Available from www.LightworksAV.com released in 1995. It has all of these characteristics (good ZPE introduction).

[24] Wilson, Jim et al. "Power from a Seething Vacuum" *Popular Mechanics*, Jan. 2002, V. 179, Issue 1, p. 22

[25] Pinto, F. "Engine Cycle of an Optically Controlled Vacuum Energy Transducer," *Physical Review B*, V.60, N. 21, 14,740,1999

[26] Private email communication.

[27] Pinto, Fabrizio, Dr Fabrizio Pinto "Progress in Quantum Vacuum Engineering: Nanotechnology & Propulsion" Second International Conference on Future Energy, September 23-24, 2006, Integrity Research Institute, Washington DC (Proceedings and DVD)

[28] Pinto, 1999

[29] See Puthoff, H.E. "Ground State of Hydrogen as a Zero-Point-Fluctuation-Determined State," *Physical Review D*, Vol. 35, No. 10, 1987, 35,3266 for further explanation of the fundamental support that ZPE provides.

[30] Dirac, P.A.M., *Directions in Physics*, J. Wiley & Sons, 1978, p.16

[31] Capra, Fritjof, *The Tao of Physics*, Shambala, 1975, p.232, 304.

[32] Dirac, p. 17

[33] *N.Y. Times*, 1/21/97

[34] Science, 12/98

[35] Lamoreaux, *Phys. Rev. Ltrs.*, 78, 1, 97

[36] U.S. Patent #4,704,622 (View any US patent at www.uspto.gov or www.google.com/patents)

[37] *Washington Times*, May 20, 2001, p. A3

[38] Bortman, Henry. "Energy Unlimited" New Scientist, Jan. 22, 2000, p. 32 cites the three-year grant to Prof. Jordan Maclay, U. of Illinois at Chicago

[39] Puthoff, H.E. p. 3266.

[40] Dr. Randell Mills directs Blacklight Power Co. in New Jersey. He is known for explaining cold fusion by "hydrinos" which involve the lowering of the ground state of the electron orbit, according to his theory, to release energy.

[41] *New Scientist*,"Why Atoms Don't Collapse,"July, 1990

[42] Fulcher et al. "The Decay of the Vacuum," *Sci. Amer.*, 12/97

[43] *Science News*, 2/8/97

[44] Petersen, I. "Peeking inside an electron's screen." *Science News*. Vol. 151, Feb. 8, 1997, p. 89 from *Phys. Rev. Ltrs.*, 1/20/97

[45] McTaggart, Lynne, *The Field: The Quest for the Secret Force of the Universe*, Quill-HarperCollins Publishers, 2003, p.30

[46] Greene, Brian. *The Fabric of the Cosmos*, Vintage Books, 2004, p. 329-335

[47] Calle, Carlos. *Einstein for Dummies*, Wiley Publishing, 2005, p. 312

[48] King, Moray. *Tapping the Zero Point Energy*, Adventures Unlimited Press, 1995, and also, King, Moray, *Quest for Zero Point Energy*, Adventures Unlimited Press, 2001

[49] McTaggart, p.15

[50] Davidson, John. *The Secret of the Creative Vacuum: Man and the Energy Dance*, C.W. Daniel Company Ltd., 1989

[51] Johnson, Gary L. The Search for a New Energy Source, Johnson Energy Corp., 1997. Also available through IRI

[52] Von Baeyer, H.C. "Vacuum Matters" *Discover*, March, 1992, V. 13, Issue 3, p. 108

[53] Nikola Tesla addressing the Institute of Electrical Engineers, 1891

[54] Einstein, A. "On a heuristic point of view concerning the generation and conversion of quanta" Annalen der Physik 17, 1905. p. 132

[55] Planck, M., p. 642

[56] Milonni, Peter. The Quantum Vacuum. Academic Press, San Diego, 1994, p. 10

[57] Einstein, A. "Zur gegenwartigen Stand des Strahlungsproblems." Phys. Zs. 10, 1909, p. 185

[58] Milonni, p. 19

[59] Einstein, A. and O. Stern. "Einige Argumente fur die Annahme einer molekularen Agitation beim absoluten Nullpunkt." Ann. d. Phys. 40, 1913, p. 551

[60] Einstein, A. "Zur Quantentheorie der Strahlung." Phys. Zs. 18, 1917, p. 121

[61] Dirac, P. A. M. "The Quantum Theory of the Emission and Absorption of Radiation." Proc. Roy. Soc. Lond. A 114, 1927, p. 243

[62] Dirac, P. A. M. "The Quantum Theory of the Electron." Proc. Roy. Soc. Lond. A117, 1928, p. 610

[63] Debye, P. "Interferenz von Rontgenstrahlen und Warmebewegung." Ann. d. Phys. 43, 1914, p. 49

[64] Wu, T. Y. The Physical and Philosophical Nature of the Foundation of Modern Physics. Linking Pub. Co., Taiwan, 1975, p. 33

[65] Pauling, L. and E. B. Wilson, Introduction to Quantum Mechanics. McGraw-Hill, NY, 1935, p. 74

[66] Pauli, Wolfgang. Selected Topics in Field Quantization. Dover Pub., NY, 1973, p. 3

[67] Snow, T. P. and J. M. Shull. Physics, West Pub. Co., St. Paul, 1986, p. 817

[68] Halliday, D., and R. Resnick, Physics Part II. John Wiley & Sons, NY, 1967, p. 1184

[69] Einstein, A., 1917, p.121

[70] Snow et al., p. 877

[71] Isaev, p. 4

[72] Ibid., p. 33

[73] Casimir, H. B. G. "On the attraction between two perfectly conducting plates."Proc. K. Ned. Akad. Wet. 51, 1948, p. 793

[74] Sparnaaay, M.J. "Measurements of Attractive Forces between Flat Plates," Physica (Utrecht). V. 24, 1958, p. 751

[75] Lamoreaux, S. K. "Demonstration of the Casimir force in the 0.6 to 6 μm range."Phys. Rev. Lett. 78, 5, 1997, p. 1

[76] Browne, Malcolm. "Physicists confirm power of nothing, measuring force of quantum 'foam.'" The New York Times. January 21, 1997, p. C1

[77] Milonni, p. 275

[78] Puthoff, Harold. "Ground State of Hydrogen as a Zero-Point Fluctuation-Determined State." Phys. Rev. D 35, 1987, p. 3266

[79] Milonni, p. 81

[80] Isaev, p. 15

[81] Milonni, p. 83

[82] Hawton, Margaret. "One-photon operators and the role of vacuum fluctuations in the Casimir force." Phys. Rev. A. 50, 2, 1994, p. 1057

[83] Ibid., p. 1057

[84] Milonni, p. 80

[85] Forward, 1984, p. 1700

[86] Iacopini, E. "Casimir effect at macroscopic distances." Phys. Rev. A. 48, 1, 1993, p. 129

[87] Haroche, S. and J. Raimond. "Cavity Quantum Electrodynamics." Scientific American. April, 1993, p. 56

[88] Weigert, Stefan. "Spatial squeezing of the vacuum and the Casimir effect." Phys. Lett. A. 214, 1996, p. 215

[89] Lambrecht, Astrid, and Marc-Thierry Jaekel, Serge Reynaud. "The Casimir force for passive mirrors." Phys. Lett. A. 225, 1997, p. 193

[90] Cougo-Pinto, M. V. "Bosonic Casimir effect in external magnetic field." J. Phys. A: Math. Gen. V. 32, 1999, p. 4457

[91] Pinto, F. "Engine cycle of an optically controlled vacuum energy transducer." Phys. Rev. B. V. 60, No. 21, 1999, p. 14740

[92] Ibid., p. 14740

[93] Valone, Thomas, *Practical Conversion of Zero Point Energy: Feasibility Study of Zero Point Energy Extraction from the Quantum Vacuum for the Performance of Useful Work*, Integrity Research Institute, Revised Edition with additional Vacuum Engineer's Toolkit, 2005

[94] Liu, Z. and L. Zeng, P. Liu. "Virtual-photon tunnel effect and quantum noise in a one-atom micromaser." Phys. Lett. A. V. 217, 1996, p. 219

[95] Valone, Thomas. "Inside Zero Point Energy." Journal of New Energy. Vol. 5, No. 4, Spring, 2001, p. 141

[96] Yater, Joseph. "Power conversion of energy fluctuations." Phys. Rev. A. Vol. 10, No. 4, 1974, p. 1361

[97] Yater, Joseph. "Relation of the second law of thermodynamics to the power conversion of energy fluctuations." Phys. Rev. A. Vol. 20, No. 4, 1979, p. 1614

[98] Yater, Joseph. "Rebuttal to 'Comments on "Power conversion of energy fluctuations."'" Phys. Rev. A. Vol. 20, No. 2, 1979, p. 623

[99] Astumian, R. D. "Thermodynamics and Kinetics of a Brownian Motor." Science, 276, 1997, p. 5314

[100] Barber, Bradley P., and Robert Hiller, Ritva Lofstedt, Seth Putterman, Keith Weninger, "Defining the Unknowns of Sonoluminescence." Physics Reports. 281 (2), March, 1997, p. 69

[101] Eberlein, Claudia. "Sonoluminescence as Quantum Vacuum Radiation." Phys. Rev. Lett. V. 76, No. 20, 1996, p. 3842
[102] Puthoff, Harold. "Gravity as a zero-point-fluctuation force." Physical Review A. Vol. 39, No. 5, March 1989, p. 2336

[103] Haisch, et al., 1994, p. 678

[104] Ibid., p. 690

[105] Beck, Christian and Michael Mackey, Astrophysics preprint, "Has Dark Energy Been Measured in the Lab?" http://xxx.arxiv.org/abs/astro-ph/0406504 June 23, 2004

[106] See for example Physics World, "Dark Energy," http://physicsweb.org/article/world/17/5/7

[107] Beck, Christian and Michael Mackey, Astrophysics preprint, "Has Dark Energy Been Measured in the Lab?" http://xxx.arxiv.org/abs/astro-ph/0406504 June 23, 2004

[108] Casimir, H.B.G. "The Casimir Effect Fifty Years Later" *Proceedings of the Fourth Workshop on Quantum Field Theory Under the Influence of External Conditions*, Michael Bordag, ed., World Scientific, Singapore, 1999, p. 3-7

[109] http://www.nasa.gov/centers/glenn/research/warp/possible.html

[110] MEC (Magnetic Energy Converter) is represented by Energy & Propulsion Systems, LLC www.ep-systems.net where one can call or email to receive the Business Plan. Roschin and Godin are the two inventors, who have several publications posted on the Internet, including the IRI website.

[111] Millis, Marc G. "Responding to Mechanical Antigravity" AIAA 2006-4913, 42nd AIAA/ASME/SAE/ASEE Joint Propulsion Conference & Exhibit, 9-12 July 2006, Sacramento, California

[112] Valone, Thomas. *Bioelectromagnetic Healing*, Integrity Research Institute, 2004

[113] Barber, Bradley P., and Robert Hiller, Ritva Lofstedt, Seth Putterman, Keith Weninger, "Defining the Unknowns of Sonoluminescence." Physics Reports. 281 (2), March, 1997, p. 69

[114] Liberati, S., and M. Visser, F. Belgiorno, D. Sciama. "Sonoluminescence as a QED vacuum effect: probing Schwinger's proposal." J. Phys. A: Math. Gen. 33, 2000, p. 2251

[115] Eberlein, Claudia. "Sonoluminescence as Quantum Vacuum Radiation." Phys. Rev. Lett. V. 76, No. 20, 1996, p. 3842

[116] Valone, 2005

[117] Mead, Frank et al. US Patent #5,590,031, col. 7, line 66. Note: patent copies can be obtained from www.uspto.gov or even better and more easily from www.google.com/patents

---

[118] George, Russ, "Catalytic Energy Science" *Proceedings of the Second International Conference on Future Energy*, Integrity Research Institute, September 22-24, 2006 (see also his DVD presentation)

[119] Mittleman, Dehmelt, and Kim. *Physical Review Letters*, V. 75, 1995, p. 2839. Also see *Physics Today*, Nov. 1995, News Page

[120] Forward, Robert. "An Introductory Tutuorial on the Quantum Mechanical Zero Temperature Electromagnetic Fluctuations of the Vacuum." Mass Modification Experiment Definition Study. Phillips Laboratory Report #PL-TR 96-3004, 1996

[121] Tsormpatzoglou, A. "Low-frequency noise spectroscopy in Au/n-GaAs Schottky diodes with InAs quantum dots" *Applied Physics Letters*, V. 87, 163109, 2005

[122] Su, Y.K. "1/f Noise and Specific Detectivity of HgCdTe Photodiodes Passivated with ZnS-CdS Films" *J. of Quantum Electronics*, V. 35, No. 54, May, 1999, p. 751

[123] Eng, Sverre T. "Low Noise Microwave Diode" US Patent #3,262,029, col. 1, line 52

[124] Johnson, J. B. "Thermal Agitation of Electricity in Conductors." Phys. Rev. 32, 1928, p. 97

[125] Callen, H. B. and T. A. Welton. "Irreversibility and Generalized Noise." Phys. Rev. V.83, 1951, p. 34

[126] Milonni, Peter. The Quantum Vacuum. Academic Press, San Diego, 1994

[127] Milonni, p. 54

[128] Milonni, p. 198

[129] Davis, E. W. et al. "Review of Experimental Concepts for Studying the Quantum Vacuum" Space Technology and Applications International Forum—STAIF 2006, edited by M.S. El-Genk, p. 1390

[130] Blanco, R., França, H. M., Santos, E., and Sponchiado, R. C., "Radiative noise in circuits with inductance," *Phys. Lett. A* **282,** 349-356 (2001).

[131] Koch, R. H., Van Harlingen, D. J., and Clarke, J., "Quantum-Noise Theory for the Resistively Shunted Josephson Junction,"*Phys. Rev. Lett.* **45**, 2132-2135 (1980).
Koch, R. H., Van Harlingen, D. J., and Clarke, J., "Observation of Zero-Point Fluctuations in a Resistively Shunted Josephson Tunnel Junction," *Phys. Rev. Lett.* **47,** 1216-1219 (1981).
Koch, R. H., Van Harlingen, D. J., and Clarke, J., "Measurements of quantum noise in resistively shunted Josephson junctions,"*Phys. Rev. B* **26**, 74-87 (1982).

[132] Beck, Christian. "Could dark energy be measured in the lab?" Astrophysics preprint, http://xxx.arxiv.org/abs/astro-ph/0406504 Nov. 24, 2004

[133] Beck, p. 2

[134] Turner, Michael S. "Dark Energy: Just What Theorists Ordered" Physics Today, April, 2003, p. 10

[135] Dobrzanski, Lech. "Low-frequency noise and charge transport in light-emitting diodes with quantum dots" J. App. Phys. V.96, N.8, Oct. 2004

[136] van der Ziel, A. "Measuring the quantum correction at zero bias in metal-oxide-metal diode noise" IEEE Journal of Quantum Electronics, Jan, 1982, V. 18, p. 17

[137] Bulsura, Adi. "Tuning in to Noise" Physics Today, March, 1996, p. 39

[138] Ibarra-Bracamontes, et al. "Stochastic ratchets with colored thermal noise" Physical Review E, Vol. 56, No. 4, October, 1997, p. 4048

[139] Iacopini, E. "Casimir effect at macroscopic distances." Phys. Rev. A. 48, 1,
1993, p. 129

[140] Haroche, S. and J. Raimond. "Cavity Quantum Electrodynamics." *Scientific American*. April, 1993, p. 56

[141] Weigert, Stefan. "Spatial squeezing of the vacuum and the Casimir effect." Phys. Lett. A. 214, 1996, p. 215

[142] Lambrecht, Astrid, and Marc-Thierry Jaekel, Serge Reynaud. "The Casimir force for passive mirrors." Phys. Lett. A. 225, 1997, p. 193

[143] Cougo-Pinto, M. V. "Bosonic Casimir effect in external magnetic field." J. Phys. A: Math. Gen. V. 32, 1999, p. 4457

[144] Forward, Robert. "An Introductory Tutuorial on the Quantum Mechanical Zero Temperature Electromagnetic Fluctuations of the Vacuum." Mass Modification Experiment Definition Study. Phillips Laboratory Report #PL-TR 96-3004, 1996, p. 3

[145]. Pinto, Fabrizio, Dr Fabrizio Pinto "Progress in Quantum Vacuum Engineering: Nanotechnology & Propulsion" Second International Conference on Future Energy, Keynote Address, September 23-24, 2007, Integrity Research Institute, Washington DC (DVD)

[146] Pinto, F. "Engine cycle of an optically controlled vacuum energy transducer" *Physical Review B*, Vol. 60, No. 21, 1999, p. 14743

[147] Ibid., p. 14743

[148] Liu, Z. and L. Zeng, P. Liu. "Virtual-photon tunnel effect and quantum noise in a one-atom micromaser." *Phys. Lett. A.* V. 217, 1996, p. 219

[149] Cheng, H. "The Casimir energy for a rectangular cavity at finite temperature" *J. Phys. A: Math Gen.* Vol. 35, March 8, 2002, p. 2205

[150] Klimchitskaya, G. "Problems with the Thermal Casimir Force between Real Metals" *Casimir Forces Workshop: Recent Developments in Experiment and Theory*, Harvard University, November 14-16, 2002, p.1

[151] Davis, E. W. et al. "Review of Experimental Concepts for Studying the Quantum Vacuum" *Proceedings Space Technology and Applications International Forum—STAIF 2006*, edited by M.S. El-Genk, p. 1396

[152] Van den Broeck, Brownian Refrigerator, *Phys. Rev. Letters*, V. 96, 210601, 2006

[153] Dhirani, et al., Self-assembled molecular rectifiers, *J. Chem Phys*. 106, (12), 22, March, 1997, p. 5249

[154] See US patent 6,635,907 "Type II Interband Heterostructure Backward Diodes" and also US patent 6,870,417 "Circuit for Loss-Less Diode Equivalent"

[155] Lynch, Jonathan et al. "Unamplified Direct Detection Sensor for Passive Millimeter Wave Imaging" Passive Millimeter-Wave Imaging Technology IX, edited by Roger Appleby, Proc. of SPIE, V. 6211, 621101, 2006
Also see: Schulman et al. "Sb-heterostructure interband backward diodes" IEEE Electron Device Letters 21, 2000, p. 353-355

[156] Young, A.C. et al. "Semimetal-semiconductor rectifiers for sensitive room-temperature microwave detectors", *App. Phys. Letters*, V. 87, 2005, p.163506

[157] Brenning et al., *J. Appl. Phys.* 100, 114321, 2006

[158] Forward, Robert. "Mass Modification Experiment Definition Study" PL-TR-96-3004, Phillips Laboratory, Air Force Materiel Command, Edward AFB, February, 1996, p. 22

[159] Fulcher and Rafelski. "The Decay of the Vacuum" Scientific American, Dec. 1979, p. 153

[160] Graneau, Peter et al. "Arc-liberated chemical energy exceeds electrical input energy" J. Plasma Physics, Vol. 63, part 2, 2000, p. 115

[161] Schwartz, Steven. "2050 Project" (see IRI Future Energy eNews archive and Steven Schwartz website) – the majority of 5000 people interviewed see small boxes in homes and offices supplying all heat and electricity by 2050.

[162] Alcubierre, Miguel. "The warp drive: hyper-fast travel within general relativity" 1994 Journal Classical and Quantum Gravity. 11 L73-L77

[163] Haisch, Bernard, and Alfonso Rueda, Harold Puthoff. "Inertia as a zero-point-field Lorentz force." Physical Review A. Vol. 49, No. 2, Feb., 1994, p. 678

[164] Clarke, Arthur C. 3001, The Final Odyssey. Ballantine Books, NY, 1997, p. 245

[165] Pinto, Fabrizio. "Progress in Quantum Vacuum Engineering: Nanotechnology & Propulsion" Second Conference on Future Energy, 2006, DVD ($20) from Integrity Research Institute, 5020 Sunnyside Ave. Suite 209, Beltsville, MD 20705 or online at www.IntegrityResearchInstitute.org

[166] Feigel, A. "Quantum vacuum contribution to the momentum of dielectric media." Physical Review Letters, Vol. **92**, p. 020404, 2004

[167] Puthoff, Harold, and S. R. Little, M. Ibison. "Engineering the zero-point field and polarizable vacuum for interstellar flight." Journal of the British Interplanetary Society. Vol. 55, 2002, p.137

[168] Ibid., p. 137

[169] Rueda, A. and Bernard Haisch. "Electromagnetic Zero Point Field as Active Energy Source in the Intergalactic Medium." 35th AIAA/ASME/SAE/ASEE Joint Propulsion Conference. June 20, 1999, AIAA paper #99-2145, p. 1

[170] Ibid., p. 4

[171] Pinto, F., ."Engine cycle of an optically controlled vacuum energy transducer" Physical Review B, Vol. 60, No. 21, 1999, p. 14740

[172] Parsegian, V. Adrian. *Van der Waals Forces: A Handbook for Biologists, Chemists, Engineers, and Physicists*, Cambridge University Press, 2006, p. 10

[173] Pinto, 2006

[174] Pinto, Fabrizio. "Apparatus Comprising of Propulsion System" US Patent Application #2006/0027709, Feb. 9, 2006

[175] Pinto, 2006

[176] Haisch, Bernard, and Alfonso Rueda, Harold Puthoff. "Inertia as a zero-point-field Lorentz force." Physical Review A. Vol. 49, No. 2, Feb., 1994, p. 678. See also, Puthoff, Harold. "Gravity as a zero-point-fluctuation force." Physical Review A. Vol. 39, No. 5, March 1989, p. 2336

[177] Froning, H.D. and R.L. Roach "Preliminary simulations of vehicle interactions with the quantum vacuum by fluid dynamic approximations" Proceedings of 38th AIAA/ASME/SAE/ASEE Joint Propulsion Conference, July, 2002, AIAA-2002-3925, p. 52236

[178] Pinto, Fabrizio. "Progress in Quantum Vacuum Engineering Propulsion" *JBIS*, Vol. 59, 2006, p. 247

[179] Calloni, E. et al. "Gravitational Effects on a Rigid Casimir Cavity" *Int. J. Mod. Phys. A*, V. 17, 2002, p. 804
Calloni, E. et al. "Vacuum Fluctuation Force on a Rigid Casimir Cavity in a Gravitational Field" *Phys. Lett. A.*, V. 297, 2002, p. 328

[180] Puthoff, H. E. "Space Propulsion: Can Empty Space Itself Provide a Solution?" *Ad Astra*, Jan/Feb. 1997, p. 42 and also see Puthoff, H.E. "SETI, the Velocity of Light Limitation, and the Alcubierre Warp Drive: An Integrating Overview" *Physics Essays*, V. 9, N. 1, 1996, p. 156

[181] Rueda and Haisch, p. 4

[182] ..."ZPE in '12" Aviation Week & Space Technology, 3/1/04, V. 160, Issue 9

[183] Obousy, Richard K. "Concepts in Advanced Field Propulsion." University of Leicester, Birmingham, Department of Physics, Lecture 5, Sec. 5.8, 1999, p.14

[184] Private email.

[185] Deffeyes, Kenneth. Hubbert's Peak: The Impending World Oil Shortage, Princeton University Press, Princeton, 2001, p. 1

[186] Greer, Steven. "Disclosure: Implications for the Environment, World Peace, World Poverty and the Human Future." Disclosure Project Briefing Document, The Disclosure Project, April, 2001, p. 2

[187] Clarke, Arthur C. 3001, The Final Odyssey. Ballantine Books, NY, 1997, p. 245

[188] Heisenberg, Werner. *Physics and Philosophy: The Revolution in Modern Science*, Harper & Row, 1962

[189] Ibid., p. 202 (Also see *Physics and Beyond: Encounters and Conversations*, Werner Heisenberg, Harper and Row, 1972)

[190] Wolf, Fred Alan. *Star Wave: Mind, Consciousness and Quantum Physics*, Collier MacMillan, 1986 (also see Taking the Quantum Leap by F. A. Wolf)

[191] Pelletier, Kenneth. *Toward a Science of Consciousness*, Dell Pub., 1978 (Chapter 2 is entitled, "Quantum Physics and Consciousness")

[192] Wolf, Fred Alan. 1986

[193] Nadeau, Robert and Menas Kafatos. *The Non-Local Universe: The New Physics and Matters of the Mind*, Oxford University Press, 1999

[194] Lindley, David. *Where Does the Weirdness Go? Why Quantum Mechanics is Strange but Not as Strange as You Think*, HarperCollins, 1996, p. 139

[195] Nadeau, p. 79

[196] Greene, Brian, *The Fabric of the Cosmos: Space, Time, and the Texture of Reality*, Vintage Books, 2005, p. 114

[197] Cowan, Daniel A. *Mind Underlies Spacetime*, Joseph Pub., 1975

[198] Friedman, Norman. *The Hidden Domain: Home of the Quantum Wave Function, Nature's Creative Source*, The Woodbridge Group, 1997, p. 121 (quoting a book by Renee Weber, *Dialogues with Scientists and Sages: The Search for Unity*, Routledge & Kegan Paul, London, 1986, p. 114)

[199] Ibid., p. 124

[200] Chown, Marcus. "Where Mind Meets Matter" *New Scientist*, August 19, 2006, p. 47 – also see his book, *The Quantum Zoo*

[201] Maclay, J. "The relationship of the human energy field to the zero point vibrations of the electromagnetic field" Proc. of Electro 78, SS/8 p. 1-12, IEEE Electronic Show and Convention, Boston, 1978

[202] LaViolette, Paul. *Subquantum Kinetics: A Systems Approach to Physics and Cosmology*, Starlane Publications, p. 29

[203] Swanson, Claude. *The Synchronized Universe: New Science of the Paranormal*, Poseidia Press, 2003

[204] Wu, Xutian et al. *Yan Xin Qigong Collectanea*, Volume Nine, International Qan Xin Qigong Association, 1997

[205] Valone, Thomas. "The Implications of the Backster Effect: An Appeal to the Scientific Community" *Proc. of Conf on New Energy*, 1993, (available from IRI)

[206] Stone, Robert B. *The Secret Life of Your Cells* , Whitford Press, 1989

[207] Yogananda, Paramahansa. *The Second Coming of Christ: The Resurrection of the Christ Within You*, Volume I, Self-Realization Fellowship, p. 21 www.yogananda-srf.org

[208] Haisch, Bernard. *The God Theory: Universes, Zero-Point Fields, and What's Behind it All*, Red Wheel/Weiser, 2006 www.thegodtheory.com

[209] Weiss, Peter. "Force from empty space drives a machine" Science News, Vol. 159, Feb. 10, 2001, p.86

[210] R.S. Decca, D. Lopez, E. Fishbach, D.E. Krause, *Phys. Rev. Lett.*, 91, (2003), 050402

[211] Boyer, Timothy. "Random electrodynamics: The theory of classical electrodynamics with classical electromagnetic zero-point radiation" *Phys. Rev. D*, Vol. 11, No. 4, 1975, p. 790. See also: Boyer, Timothy. "General connection between random electrodynamics and quantum electrodynamics for free electromagnetic fields and for dipole oscillator systems" *Phys. Rev. D*, V. 11, N. 4, 1975, p. 809 and also: Boyer, T. H. "Quantum Zero-Point Energy and Long-Range Forces" *Ann d. Phys.* (New York) V.56, 1970, p. 474

[212] Lamoreaux, S. K. "Demonstration of the Casimir force in the 0.6 to 6 μm range."Phys. Rev. Lett. 78, 5, 1997, p.2

[213] Forward, Robert. "Extracting electrical energy from the vacuum by cohesion of charged foliated conductors." Phys. Rev. B. 30, 4, 1984, p.1701

[214] Milonni, Peter. The Quantum Vacuum. Academic Press, San Diego, 1994, p. 111

[215] Casimir, H. B. G. "On the attraction between two perfectly conducting plates."Proc. K. Ned. Akad. Wet. 51, 1948, p. 793

[216] Boyer, 1980, p. 66

[217] Milonni, 1994, p. 19

[218] Puthoff, Harold. "Gravity as a zero-point-fluctuation force." Physical Review A. Vol. 39, No. 5, March 1989, p. 2336

[219] Puthoff, 1987, p. 3266

[220] Haisch, Bernard, and Alfonso Rueda, Harold Puthoff. "Inertia as a zero-point-field Lorentz force." Physical Review A. Vol. 49, No. 2, Feb., 1994, p. 678

[221] Lamb, 1947, p. 241

[222] Baym, Gordon. Lectures on Quantum Mechanics. Benjamin/Cummings, Reading, 1978 p. 99

[223] Planck, M. "Uber die Begrundung des Gesetzes der scwarzen Strahlung." Ann. d. Phys. 37, 1912, p. 642

[224] Callen, H. B. and T. A. Welton. "Irreversibility and Generalized Noise." Phys. Rev. _83, 1951, p. 34

[225] Eberlein, Claudia. "Sonoluminescence as Quantum Vacuum Radiation." Phys. Rev. Lett. V. 76, No. 20, 1996, p. 3842

[226] Milonni, 1994, p. 111

[227] Baym, p. 126

[228] Milonni, 1994, p. 99

[229] Ford, L.H. et al. "Focusing Vacuum Fluctuations" Casimir Forces Workshop: Recent Developments in Experiment and Theory, Harvard University, November 14-16, 2002, p. 1

[230] Ibid., p. 19

[231] Milonni, p. 200

[232] Weigert, Stefan. "Spatial squeezing of the vacuum and the Casimir effect."Phys. Lett. A. 214, 1996, p. 215

[233] Hu, Z. et al. "Squeezed Phonon States: Modulating Quantum Fluctuations of Atomic Displacements" Phys. Rev. Lett. V. 76, 1996, p. 2294

[234] Wiegert, p. 217

[235] Weigert, p. 219

[236] Dodonov, V.V. et al. "Squeezing and photon distribution in a vibrating cavity" J. Phys. A: Math Gen. V. 32, 1999, p. 6721

[237] Maclay, J. "Unusual properties of conductive rectangular cavities in the zero point electromagnetic field: resolving Forward's Casimir energy extraction cycle paradox" Proceedings of Space Technology and Applications International Forum (STAIF), Albuquerque, NM, January, 1999, p. 1-5

[238] Marachevsky, V.N. Modern Physics Letters A, V. 16, 2001, p. 1007

[239] Maclay, 1999, p. 6

[240] Sagan, C. Cosmos, Random House, New York, 1980, p. 37

[241] Zheng, M-S., et al. "Influence of combination of Casimir force and residual stress on the behaviour of micro- and nano-electromechanical systems" Chinese Physics Letters, V. 19, No. 6, 2002, p. 832. Also reviewed in PRACTICAL CONVERSION OF ZERO-POINT ENERGY: Feasibility Study Of Zero-Point Energy Extraction From The Quantum Vacuum For The Performance Of Useful Work, by Thomas Valone, Integrity Research Institute publishers, 2005 edition, p. 49

[242] Cole, Daniel, and Harold Puthoff. "Extracting energy and heat from the vacuum." Physical Review E. Vol. 48, No. 2, August, 1993, p. 1562

[243] Ibid., p. 1563

[244] Budker and Corsini, UC Berkeley, Physics Dept., 2006 http://socrates.berkeley.edu/~budker/Physics250_Spring0607/Corsini_Casimir.ppt

[245] Steitwieser, Andrew et al. Introduction to Organic Chemistry, Second Edition, Macmillan Publishing Co., 1981, p. 294

[246] Popov, A.A. "Long throat of a wormhole created from vacuum fluctuations" Class. Quantum Grav. V. 22, 2005, p. 5223

[247] Milonni, p. 184

[248] Parsegian, V. Adrian. *Van der Waals Forces: A Handbook for Biologists, Chemists, Engineers, and Physicists*, Cambridge University Press, 2006

[249] Professor Chris Binns (Physics and Astronomy), University of Leicester, eBulletin, 2005, on his exciting project connected to the 'zero-point energy' of space: http://ebulletin.le.ac.uk/features/2000-2009/2005/08/nparticle-82w-fqr-2cd

[250] Forward, 1984, p. 1700

[251] Haroche, S. and J. Raimond. "Cavity Quantum Electrodynamics." Scientific American. April, 1993, p. 56

[252] Weigert, Stefan. "Spatial squeezing of the vacuum and the Casimir effect."

Phys. Lett. A. 214, 1996, p. 215

[253] Vacuum Engineer's Toolkit is available for free online at www.IntegrityResearchInstitute.org and in the book, *Practical Conversion of Zero Point Energy* by T. Valone
[254] Scully, M.O. et al. "Extracting work from a single heat bath via vanishing quantum coherence" Science, Vol. 299, Issue 5608, 2003, p. 862

[255] Milonni, P.W. "Photon Steam Engines" Physics World, April, 2003, p. 2

[256] Ibid., p. 3

[257] Allahverdyan, A.E. and T.M. Nieuwenhuizen "Extraction of work from a single thermal bath in the quantum regime" Physical Review Letters, vol. 85, No. 9, August, 2000, p. 1799

[258] Ibid., p. 1800

[259] Prigogine, Ilya. *Order Out of Chaos: Man's New Dialogue with Nature*, Bantam Books, 1984

[260] Prigogine, p. 286

[261] Prigogine, p. 276 (He references two of his own works here: Courbage, M. and I. Prigogine. "Intrinsic Randomness and Intrinsic Irreversibilty in Classical Dynamical Systems" *Proc. of the Nat. Acad. Of Sci.*, V. 80, April, 1983 and also, Prigogine et al., "The Second Law as a Selection Principle: The Microscopic Theory of Dissipative Processes in Quantum Systems" *Proc. of the Nat. Acad. Of Sci.*, Vol. 80, 1983, p. 4590)

[262] Prigogine, p. 181

[263] Baym, Gordon. Lectures on Quantum Mechanics. Benjamin/Cummings, Reading, 1978, p. 126

[264] Joos, Georg. Theoretical Physics. Dover, NY, 1986, p. 743

Review

Zero Point Energy: The Fuel of the Future by Thomas Valone, Integrity Research Institute, 2007, 236 pages, $22, ISBN 978-09641070-2-1

How to Use Zero Point Energy

Zero bias diodes are one of the many overlooked converters of vacuum fluctuations for the production of DC electricity that are explained in detail in a new zero point energy breakthrough technology book paving a path toward carbonless, emission-free energy

UP UNTIL now the use of zero point energy (ZPE) for electricity generation was mere fantasy and science fiction. No one thought it possible for ZPE to offer a source of unlimited energy for homes, cars, and space travel, even though it is agreed that the energy density of the vacuum *exceeds the energy density of matter.* Some "experts" still say zero point

"You have to appreciate that looking in the noise level is where ZPE manifests"

energy can do nothing useful. The experimental evidence however, incontrovertibly demonstrates that ZPE-caused fluctuations in electricity have been measured in tunnel diodes, coils and diode rectifiers. Furthermore, ZPE can be focused, polarized, amplified, turned on or off, attract or repel matter in one direction, move dielectric fluid and even electrons for a desired electricity flow in a circuit.

Max Planck's 1912 Second Law proved that ZPE was useful

The details of these discoveries are presented in the book, *Zero Point Energy: The Fuel of the Future*, which concludes that there is a major future **energy trend** in the physics community toward ZPE power. For example, there exists a class of diodes that operate at "zero-bias" (no voltage applied to make them work) and up into microwave frequencies, which are suitable for generating trickle currents from the zero point energy quantum vacuum because of natural nonthermal electrical fluctuations (1/f noise) which have ZPE as their source. Thus, in this age of "energy harvesting" for various nanotechnology circuits, the all-pervasive ZPE may be the next big thing to start harvesting, according to Valone.

One class of ZPE diodes that are interesting function by "tunneling," even at zero voltage (zero bias). Several microwave diodes today exhibit this feature. However, "You have to appreciate that looking in the noise level is where ZPE manifests," says author Thomas Valone. "**Nature has been helpful** since 1/f noise in the diode is generated at the junction itself and therefore, requires no minimum signal to initiate the conduction in one direction." Josephson junctions, metal-oxide-metal diodes and other semiconductors demonstrate substantive generation of energy from ZPE.

A suitable zero bias diode includes the DARPA developed night vision "Unamplified Direct Detection Sensor" (Proc. of SPIE, V. 6211, 621101, 2006). Also, in a series of experiments from 1980 to 1982, Koch measured zero-point voltage fluctuations in Josephson tunnel junctions (Phys. Rev. Lett., V. 47, 1216, 1981). The Koch result is striking confirmation of the reality of ZPE and suggests that the vacuum fluctuations can do real work (creating electricity). Koch paved the way for other ZPE circuit discoveries such as Blanco who reports amplified ZPE noise with large inductive coils (Phys. Lett. A, V. 282, 349, 2001).

Eric Davis, from the Institute for Advanced Studies in Austin, Texas reported at STAIF-06 negotiations with Lockheed Martin to fund a replication of Blanco's work. Professor Christian Beck (Queens College, London), author of a textbook on the vacuum fluctuations of ZPE, published a paper in 2004, "Could Dark Energy be Measured in the Lab?" (http://xxx.arxiv.org/abs/astro-ph/0406504) based on work by Koch.

All of these developments make exciting reading in a book **full of pictures, history, science and suspense** on a subject that has not been accessible to the general public in any digestible form until now.

Reviewers say "...his book is indeed welcome" -- Jeane Manning, author of *The Coming Energy Revolution* (Avery). "Valone describes with masterful ease, and in an easy-to-understand timeline, the history and science underpinning this esoteric, yet vitally important subject - one that is rightly beginning to move from left- to centre-stage in our understanding of the universe and, indeed, the very essence of reality, both at a macro and quantum level" -- Nick Cook, former Science Writer, ***Janes Defence Weekly*** (UK) and author of *The Hunt for Zero Point* (Random House).

Zero Point Energy, The Fuel of the Future, being so comprehensive and yet readable, may yet become a classic. A popular Valone ZPE conference lecture has also found its way onto YouTube (http://www.youtube.com/watch?v=UMNZyOs7UU0). More information is available at www.amazon.com ●

Author Profile

Thomas Valone, PhD, PE did his thesis on the "Practical Conversion of Zero Point Energy from the Quantum Vacuum for the Performance of Useful Work" and is a former college physics teacher. He is the author of *Harnessing the Wheelwork of Nature, Electrogravitics II, The Homopolar Handbook* and *Bioelectromagnetic Healing*.

Book Review by Moray B. King

author of ***Tapping the Zero Point Energy***
and ***Quest for Zero Point Energy***

Zero Point Energy - The Fuel of the Future
by Thomas Valone

This book can change the world. Zero-point energy is little known outside the physics community or the frontier research community. Yet here is an energy source that just might be the miracle that humanity is hoping for to free us from the present energy crisis that is starving the world. Thomas Valone's great contribution is explaining in clear layman's terms the concept and availability of zero-point energy, and how to go about tapping it as an energy source. The more people that understand zero-point energy, the more support there will be for its research. Moreover for the advance reader, the book contains abundant citations into the physics journals as well as the internet to provide specialized technical information. Valone has a clear ability to communicate to both levels of readers in a fascinating manner.

Zero-point refers to absolute zero degrees (Kelvin) temperature, where all heat and radiation is absent. Yet the completely empty fabric of space, referred to as "the physical vacuum," contains a plenum of fluctuating electricity. It is like a sea of never-ending, microscopic lightning flashes at a scale much smaller than the elementary particles (like protons, electrons). These energetic fluctuations occur everywhere and interact with all elementary particles. (It might even be the source of the particles' mass/energy where the zero-point energy continuously feeds the particle much like the flow of a stream feeds a whirlpool's existence.) What is exciting is that nanotechnology can coherently interact with the zero-point fluctuations. In one chapter Valone overviews a class of devices known as "zero-bias diodes," that effectively gate the vacuum fluctuations so that they can be absorbed into useful electricity. This is the holy grail of advance zero-point energy devices, for when scaled up with millions of gates (much like is done for computer chips), it would yield a solid state energy machine with no moving parts. What is encouraging is to read how much progress has been made toward these types of devices by the standard scientific community.

A new source of energy is not the only payoff that can come from cohering zero-point energy. Novel forms of propulsion and space travel can also manifest. Valone overviews research that shows the zero-point energy could be the basis of inertia and gravity. Controlling the zero-point energy around a craft could yield propulsion much like described for UFO's where rapid acceleration can occur without causing inertial stress for the occupants. The results of zero-point energy research can yield miraculous technologies that heretofore have only been the dreams of science fiction.

Thomas Valone is a master lecturer and master author who brings to life the exciting possibilities offered by zero-point energy research. Anyone who reads his book will be inspired, and such inspiration will help accelerate the successful research and development of new technologies. With his book Thomas Valone has made a significant contribution to uplift our world.